RECONNECTING SCIENCE AND FAITH

BRIDGES

Walter and Susan
What a pleasure to send you my book.
I hope that the views from the Bridges
within these pages will enhance the beauty
of the engagement between science + faith
for you.
Happy Holidays 2012
with affection
+ thanks
Steve

RECONNECTING SCIENCE AND FAITH

BRIDGES

STEPHEN PARKER

TATE PUBLISHING
AND ENTERPRISES, LLC

Published by Tate Publishing & Enterprises, LLC
127 E. Trade Center Terrace | Mustang, Oklahoma 73064 USA
1.888.361.9473 | www.tatepublishing.com

Tate Publishing is committed to excellence in the publishing industry. The company reflects the philosophy established by the founders, based on Psalm 68:11,
"The Lord gave the word and great was the company of those who published it."

Cover design by Joel Uber
Interior design by Erin DeMoss

Published in the United States of America

ISBN: 978-1-61862-342-3
1. Science/ Philosophy & Social Aspects
2. Religion/ Religion & Science
12.05.04

DEDICATION

For Lesley and Ray, their children, Tyler, Will and Maya. For Chris and Addison, their daughter, Gloria. For Carrie and Ryan, their children, Aidan, Ainsley and Blaithin. For Dave and Jenny, their daughter Keira. May they love science and be people of faith.

I want to thank my wife, Jeanie, for her encouragement, her reactions to ideas, initial editing and, most of all, for her patience through the process. I am in gratitude beyond measure to my parents who encouraged academics and faith and bought me my first telescope and Bible in 1952. I am grateful for my boyhood friend, Bill Gilbert, and our meanderings through the hills and valleys of religion and evolution. I am in debt as well to The College of William and Mary, its courses in Astronomy and Physics, the introduction to Philosophy and a way to refine my skepticism in Sociology. I am particularly thankful for the professors at Berkeley Divinity School at Yale who taught me how to think critically about the Bible and maintain and strengthen my faith at the same time. Perhaps, most importantly, they allowed for doubt to be mixed with faith in the development of our theological minds. I am grateful to the parishes I have served, St. Luke's Darien, as seminarian and curate, Trinity Episcopal Church, Collinsville, CT., and St. Matthew's Church, Wilton, CT as rector, St. James the Fisherman, Islamorada, Florida where I served as supply and interim minister. I have a special gratitude for the experience at Salisbury School in Connecticut where I served as Chaplain, headed the Department of Philosophy and Religion, taught astronomy and theology to seniors and coached two varsity sports. My gratitude goes as well to the parishioners at St. John's Fishers Island where during twelve summers I served as Rector. In addition to a regular theology oriented book group, I engaged a throng of stargazers in a Starry, Starry Night summer skies program.

I am particularly grateful to my friend and neighbor Jacques Linder for the encouragement and enthusiastic reading of revision upon revision. Jacques, an MIT graduate, engineer, artist and poet never let me doubt that at the right time this book would be published. Jacques was also the photographer for the book jacket and his poetry heads a couple of chapters.

Eloise Hanner, an author herself, was the first critical reader outside my inner circle of family and friends. Her sharp critique enabled me to get serious about the writing. Other friends who labored through the reading of the rough drafts were Ann and Bill Moore, and John Meyers. Thanks also to Cynthia Riley for her encouragement and for introducing me to the work of astronomer, Maria Mitchell, and to Peter Rugg whose trip to the Galapagos gave me a spiritual sense of the islands. My former colleague in ministry, Steve Jacobs, now Vice President of Regis College in Colorado, gave some critical advice and helped me reorder some of the chapters. Denny Wachter, a teaching colleague at Salisbury School, tirelessly went through the final drafts with me on the way to translating the book into Spanish. We hope to have a Spanish version of Bridges in the future.

Midway through the process, Lisa Sheble, my former colleague in the Department of Philosophy and Religion at Salisbury School, dissected, stripped, honed and debated the text with me, mostly by email and phone from Florida to Connecticut. She labored for countless hours to help me make this more readable and useful. I am forever grateful to her.

Thank you also to Hal Ellens of Edwin Mellen Publications and ABC-CLIO Praeger Press who encouraged me to expand the book and gave me some additional excellent advice, and to my first professional editor, Deborah Sosin, who challenged me again and again and went through every jot and tittle of the text. In addition to her editing skills, she found just the right touch in helping me to see the need for rewriting many sections. Finally, to the staff of Tate Publications who believed enough in the book to take it on. Particularly, I am grateful to Donna Chumley, an acquisition editor at Tate Publishing who saw the value in the book, Thomas Beard, my conceptual editor who challenged me further and honed the final two drafts, and Joel Uber, a graphics designer who produced the book cover. Lastly, thank you to Erin DeMoss who did such a fine job with the layout of the book.

TABLE OF CONTENTS

PREFACE

The world of faith and the world of science are often disconnected. It was not always so. Copernicus, Galileo, Newton, Schweitzer and Einstein were all men of faith as well as of science. This book seeks to reconnect men and women of faith to the wonder of science and men and women of science to the wonder of faith. I believe that we are at a time in our history when the world could benefit greatly from the collective force of science and faith.

As faith is used in so many different ways, let me write at the outset a simple definition of faith for me. Faith is an active response to being alive. It is not a rigid position. Its purpose is not principally to affirm doctrines or theological postures. It is a way of life that quells fear and enables wholesome living and growth. Faith invites curiosity and does not banish doubt; rather it absorbs doubt and learns from what doubt has to offer. Most importantly, faith transcends doubt; it does not have to beat down doubt. At times doubt can be an ally. Doubt by itself cannot destroy faith; only fear can harm it. Perfect love and perfect faith cast out fear. In this sense faith and love are mystically tied together. Faith that God is love is a life posture, not a position to be proved. When something is unknown, faith smiles and waits for interpretation to inform and enrich. Faith is necessary for new knowledge to be embraced. It is critical to all people who wish to live a life of hope, inspiration and love.

The thrust of this book is that, through the centuries, the body of faith and the body of science are evolutionary and complementary in nature. Men and women of faith and of science sit on the shoulders of past giants, people of profound spirituality and reason. Faith and science provide checks and balances for each other. Each needs the other, and each expands the vision of the other. Science owes faith its energy and motivation, and faith owes science its grounding in the material world.

My own background is in theology and astronomy. I am an Episcopal priest who, for part of my career, taught classes in theology, the Hebrew and Christian Scriptures and astronomy to high school seniors. In this book, I limit the discussion to the fields with which I am most familiar, recognizing that science and faith are broad subjects, and that others can use their own faith or their own understanding of science to augment my thesis.

In particular I demonstrate that the Bible, especially the Hebrew Scriptures, is a book of spiritual evolution, and that the Christian Testament feeds off of and is nurtured by the heights of Hebrew thought. The Christian Testament is, of course, only possible with the Hebrew Scriptures as its foundation. This book celebrates the development of biblical thought as much as it celebrates science's understanding of the evolution of life on the planet.

I have written this book because my faith became much richer when I welcomed science into it. The world is magnificent and full of mystery and may be celebrated and embraced more strongly when science and faith join hands.

For me, the engagement of these two disciplines has been a lifetime journey. I invite you to take this journey with me. I open the book with a fanciful engagement between an icon of faith and an icon science, Jesus and Darwin. In my imaginative exploration, the two men encounter and test each other and walk or sail in the same direction. If they can connect on a scientific and spiritual level, it remains for us to open our minds and spirits to do likewise.

GIANTS ON THE BEAGLE

According to your faith
let it be done to you. (1)

—Matthew 9:29

I think I have found out the simple way
by which species become exquisitely adapted
to various ends. (2)

—Charles Darwin, in a letter to Joseph Hooker

Somewhere in the wrinkles of time and space, Jesus of Nazareth engaged Charles Darwin at a celestial gathering. After some initial bantering, Jesus suggested that they spend more time together. Charles warily warmed to the idea as he had been a notoriously shy man, but this extraordinary opportunity and Jesus' friendly demeanor calmed his nervous spirit and overcame his initial reluctance. He suggested that Jesus accompany him on his cosmic sailing vessel, HMS Beagle. Charles proposed that they re-enact a portion of his famous trip. The two would sail to the Galapagos Islands, possibly the most auspicious place where they could ponder great ideas. It would be away from the rush of humanity and heavenly responsibilities and within Charles's comfort zone. The location was perfect as it was during the leg of the journey that Charles took between 1832 and 1835 that some of his greatest thoughts on the evolution of the species emerged.

Jesus assured Charles that he would not be a hostile passenger, and that it would be a wondrous moment in time to share each other's perspectives and revolutionary ideas. So, Captain Darwin prepared the Beagle for this journey, which mimicked only in part his initial trip to these magical islands. They would embark at a port in Guayaquil, Ecuador.

Shortly after setting sail, when they had their bearings and the open water was before them, the conversation between Charles and Jesus began.

"How are you feeling on this wide open ocean?" Charles asked Jesus.

"Just fine" said Jesus, as the wind whipped back his long, flowing hair. "I developed my sea legs on the Sea of Galilee with some of my closest friends. They were mostly fishermen, but we had plenty of talking time on the quiet waters, especially when the fish weren't biting. However, I have never been on such a vast sea as this. The great expanse I am used to is the desert, where I spent countless hours meditating and thinking about life."

At that moment the wind quickened and Charles got up to adjust the sail, reminding Jesus to be sure to duck when the boom came about. When it was calm again, Charles said, "So many of my ideas came from spending time alone, too. I had several mentors, and I am especially indebted to Charles Lyell. His book, *Principles of Geology*, changed the way I saw the world. Subsequently I enjoyed contemplating my own observations."

"You were fortunate to have mentors. I had them as well. When I returned from Egypt as a boy, I continued to discuss the Scriptures with the local temple congregation, and I especially liked it when my family travelled to Jerusalem, which allowed me to talk and listen and debate with the Pharisees in the temple areas. We spoke mostly Aramaic, though the scrolls in the temple were in Hebrew, the formal institutional language. Along with the education I received in schools and that which I sought out in the temple, I spent time every day meditating, praying to God. Gradually I began to think and feel deeply about what was needed to change the world and my role in it–but more about that later. Tell me a little about the great adventure you took to the Galapagos Islands."

"You know, I have been extraordinarily lucky. On my original voyage, we sailed down the east coast and up the west coast of South America. At one point, we anchored the Beagle off the coast of Argentina at Punta Alta, and I discovered some enormous, ancient, extinct, mammalian bones mixed with newer, smaller, recognizable ones and wondered how they were related. When we headed up the Chilean coast near Valparaiso, by coincidence, Mount Osorno erupted. Then an earthquake struck, and the shoreline dramatically receded by eight feet. I began to consider the possibility that even the Andes had at one time been under water. Think about that and what the changing world meant for the creatures that inhabited those areas." (3)

"Well, it became clear to me that great land masses that at one time were submerged under water and then emerged ages later transformed the planet. Many species on Earth have needed to adapt physically because of the Earth's dynamic alterations over time. When I pondered these monumental transformations, I came to believe that the Earth was a lot older than we used to think. I now believe that the birds and the land animals have been here for ages and ages."

The wind picked up again and Charles had to attend to the mainsail. Jesus moved to the starboard where he could assist with the line. When the wind calmed the two men relaxed again and continued their conversation.

"Charles, your discoveries and intuition about the past are amazing," said Jesus. "I would like to put your insights and discoveries in a theological perspective. Nothing that you have said should trouble those who believe that God is the author of all Creation. I am sure that the Lord of Creation is delighted with your perceptions. Discovering that which comes closer to the truth is never something to be afraid of. Truth is emancipation and is often wrought out of pain and struggle and yields itself over time."

"I am glad you feel that way. As you may know, I was training to be a cleric but felt I could make a greater contribution to the world through biology and geology."

"You were absolutely right to follow that path. You demonstrated a great gift for it."

"Thank you," said Charles. "I think we are about to embark on a shift in human consciousness that has few parallels. If what I am thinking turns out to be true, then human beings are part of an evolutionary chain. Perhaps we were not created out of dust but out of a longer, far more expansive divine plan. God is patient isn't He? Eve did not spring from Adam's rib. It appears that human beings evolved from lower primates, who in turn evolved from lesser mammals, who in turn..." Darwin paused.

Jesus thought for a moment, then replied, "You are right that your theory is altering the way human beings feel about their origin. But your theory should help support biblical study as well. I hope that it will also help restore the original intention of the writers of the early Scriptures. Those writers were adept at using stories and metaphors to express what was, at the time, inexpressible and unknown. Furthermore, those writers did not have the benefit of your knowledge and discoveries. What they really intended was for us to celebrate that God was the author of Creation and that human beings could have a special relationship to Him. Your theories do not materially change the notion of and belief in God's authorship of creation. In fact, you have enriched us with the knowledge of the magnificence of the creative process. After all, the God of this world and of all Creation is eternal, without dimension, timeless and omnipresent, and you have given us a strong taste of what that means. You have increased our awe and let us glimpse a little more of God's world and time. We can now better appreciate that the days of Creation were a metaphor for ages of time, God's time."

"I am glad you see it that way. It may make it easier for future biologists and other scientists to expound upon my fledging theory if there are more people who think as you do."

"Unfortunately, people let go of cherished positions very slowly. I know from personal experience that many did not want to hear my views of what was needed in the world. I was trying to let them in on a big secret: that God's kingdom, the one that really matters, is not the kingdom of greed, where money and power rule, where a coin of the realm determines the value of everything. Instead, it is a kingdom where one turns the other cheek rather than returns the blow, and where one's strength is measured in kindness and compassion rather than in the flexing of one's muscles. I also wanted to let them know that while God's Kingdom is eternal, it dawns or breaks in for human beings here on earth. But, you see the message was difficult for many and it got me to a hill on Calvary."

"I appreciate your viewpoint so much. You know, I had my own personal battles and demons to fight. I was not really a born salesman. I was kind of a loner. I loved the discovery process, I loved the writing, but I never wanted to be a public figure. I really didn't want to go out and defend my theories. In the end, I mostly left that to others. I was not cut out to do all that marketing."

"I know. That public marketing is the tough part," Jesus responded. "But, you thankfully did not have to do that all by yourself. Others have aided you and continue to pick up the banner for you. As for me, I needed and enjoyed my time with God in the desert. I became grounded in the meditative, communing part, particularly my private prayer vigils, and I embraced the journey involved in studying the Scriptures. The Scriptures slowly revealed to me what was the essence of faith and what was important in how human beings lived their lives. I had many people trying to dissuade me from taking too strong a stand, including my mother and my sisters and brothers. They wanted the truth, but in easy-to-digest portions. They wanted to shelter me."

Jesus paused, took some deep breaths, looked out on the vast expanse of ocean and sky and sighed. Then he continued, "Even my strongest leader, Cephas, whom you know as Peter, the rock, wanted me to take it slowly. After all, he wanted to live a nice long life in this world and he wanted me to be there with him. I had to be rather cruel to him and the other disciples and my family, telling them they were getting in the way of the revolutionary path that God wanted me to take. You, too, had a difficult row to hoe. You could not leave it entirely to others. You had to have courage, but others eventually helped to carry your banner. They did for me as well. My mission was only as successful as it inspired others to live the Gospel, the "Good News," which I taught. Your banner will later be carried by biologists, geologists and eventually by all educated people. Even churchmen will join you in time. But, during your sojourn on earth, you had to stand alone at times and be scorned. That is the way it has always been for those who dare to embark upon a new course."

Jesus looked up to watch two magnificent frigatebirds soar high above the Beagle. "Look ahead," Charles cheered! "There are the first islands of the Galapagos. They stretch out like a string of pearls from the northwest to the southeast."

Jesus gazed, and then asked, "How do you think they formed?"

"It appears that the islands rose up from the welling up of lava from magma deep beneath the Earth's surface through undersea volcanoes until they pierced the surface and rose above the sea."

Both men marveled, taking in the vista before them. The rocky shore appeared to stretch out forever; the sky seemed to enclose a new world; nearby a flock of flamingoes waded at the shore unconcerned while blue-footed boobies dove around them; and immediately, Jesus and Charles felt as though they were in a world unlike any other.

It was now time to drop anchor. Charles and Jesus worked together and prepared to disembark. When they were ready to wade ashore, Jesus said, "I am starving. Are you hungry?"

Charles laughed and said, "I sure am, and I wouldn't be surprised if we found some interesting lunch tables on the shore."

Jesus said with a twinkle in his eye, "I did bring two small fish and a barley loaf."

"Very funny, but you may not need them. There is much to eat here."

Once ashore, Jesus said, "Look at those boulders. They seem to be moving." As they drew closer, Charles smiled and remained silent.

Then he chuckled and said, "You see they are not stones at all, but the biggest tortoises one could ever imagine."

The tortoises did not move, and were obviously unafraid. Charles, always the adventurer, sidled up to one and touched it; it still did not move. Then, carefully, he sat upon it. Jesus did the same with a tortoise only a few yards away, and the two human icons started their lunch sitting upon the great turtles, contemplating their surroundings. They were silent for a considerable time, just drinking in the beautiful, primitive scenery, and then Charles looked at Jesus and each knew that the grittiest of questions and thoughts would follow. For this they moved to some lava boulders not far away.

"My imagination flows into the future," said Charles. "This is what I see, and what troubles me the most. These creatures do not seem to be created out of nothing, deus ex machina, by the hand of God, at least not all at once. All the creatures have a history, a long history. The species are connected to one another; one species is not necessarily more special than another. Each seems to adapt from his old environment and over time appears quite different from his forebears."

As the two men conversed, they watched as the two tortoises upon which they had been sitting moved and then lumbered toward the shore. There was something so primitive about them, creatures from a primordial past still inhabiting our world.

Charles, enjoying his audience, continued. "The crux of the matter is what I am going to call natural selection. It is a survival pattern. The fittest will survive in the sea, or on the land, or in the air. Some have even come out of the sea to survive on land, slowly evolving into land creatures and some developing the ability to fly. In addition to the survival aspect of evolution influenced by the climate, there is a sexual component to it as well. Since reproduction is necessary to survival, sexual competition abounds. The drive to survive over time alters the species as the environment in which the creatures live changes."

Charles looked intently, and then pondered aloud, "Can people of faith find a way to adjust to this ground-shaking concept with integrity? Can people of faith incorporate this theory of evolution into their faith perspective? Can our two ways of thinking be compatible?"

"Certainly!" Jesus said. "I believe the spiritually minded, biblically oriented person will adapt to this new environment. One will have to embrace a greater vision of the creative God and let go of a certain theological perspective that insists on building a universe of human proportions rather than godly ones. God's ways are in some measure inscrutable, and not even I understand all that the Father of Creation had in mind."

They wandered to the water's edge and Jesus continued his thought.

"It has taken a long time for evolution to come to the point where human beings live and have an awareness or intuitive sense of God. A consciousness of God is part of the evolutionary advance in human beings. It moves unevenly through the centuries, but the Spirit, God's eternal gift to us, inevitably draws this inner awareness out into the open where it can be shared. It has detours, just as your evolution among the species does, but over time, certain truths are realized. 'The Golden Rule' is one such example: 'Do unto others as you would have them do unto you.'(4) So many people over the globe in many different religions, or those with no religion at all, recognize the power and goodness of this admonition."

"I hope you are right, for there does not seem to be much to be gained by turning back. The world needs to learn and adjust. This adjustment may be part of the evolutionary process itself."

"I like the way you said that. We have arrived at a point in human evolution where we may influence evolution itself in its broader context dramatically. In the future, it will not only be the environment or sexual competition that creates new beings, but human evolution may also proceed within the human mind and heart. We can imagine a better world and produce structures to change it. We can even alter our behavior by embracing justice and compassion. We can dramatically affect evolution in the future, but how we do it is very important and will determine everything about life on the planet. If we want power, we can change the world to reflect this value. If we want a world where the love of God and the love of one's neighbor are paramount, we can make that happen."

Charles and Jesus then fell silent, each with his own thoughts. Each had a peaceful-looking smile. It was a special time—Charles Darwin, the naturalist, and Jesus, the Galilean teacher, enjoying an unprecedented moment. The two men boarded the Beagle once again and put out to sea. They had forged a friendship and any future sailing would be inspiring entertainment. There was much more to share.

Jesus said, "Charles, if the wrinkles in time and space will again permit it, let us meet again. I would enjoy it very much."

"Great. Where do you propose we go?"

Jesus said, "Let's meet at a port near the ancient town of Capernaum, on the northwestern shore of the Sea of Galilee. I would like to sail with you there; it was an area I used to call home, and I think you'll enjoy it."

"Sounds like a wonderful idea. I'll be there at the appointed time," said Charles.

COMING OF AGE IN SCIENCE AND FAITH

For you are my hope, O Lord God,
my confidence since I was young.

—Psalm 71:5

Science is an ongoing process. It never ends. There is no single ultimate truth to be achieved, after which all the scientists can retire. And because this is so, the world is far more interesting.... (5)

—Carl Sagan, Cosmos

As a twelve-year-old boy, the nighttime sky drew me into its web of wonders. It was on clear Massachusetts winter nights that my love affair between science and faith was born. At my urging, my parents purchased a 2.4-inch Unitron refractor, a 1952 Christmas present. I didn't realize it at the time but that small telescope was about the same strength and caliber as the one Galileo used to change our perception of the relationship of the earth to the solar system. An adventure was born, one that took me to the distant heavens to stretch my understanding of the God of Creation and our place in God's cosmos. I had already begun another adventure that was rooted in the Earth, a spiritual path led by the man from Galilee who walked in the flesh so long ago. As my life proceeded, these two paths wove into a greater highway and a bigger adventure.

Active interest in astronomy or in worlds beyond the moon was rare in suburbia in the early 1950's. I knew of no one in my end of town who either owned a telescope or studied the night sky. One cold evening in 1953, I was out near the street where I lived working with my new telescope, peering through the trees for a look at Mars. A police cruiser came by, slowed and stopped; an officer wanted to know what I was up to. Enthusiastically, I asked him if he'd like a look at the red planet. He zoomed away as if he were in danger of catching a virus.

At about the same time that I was viewing the four largest moons of Jupiter and barely making out some faint markings on Mars, I was preparing for confirmation in the Episcopal Church. My best friend, Bill Gilbert, and I frequently walked home at the end of Sunday's class, and before long the subject of our conversation turned to creation and evolution. Could God have been the author of each? We dared to ask a question not easily discussed at home; did God create us or did we create God? Bill, my cohort in adolescent theological meanderings, grew up to become a college professor, a zoologist as well as a Christian apologist, and I became an Episcopal priest.

My parents, who were happily married for 58 years, helped to strengthen my appreciation of science and faith. My mother was a bright, Vassar-educated psychology major, and a Manhattan-bred Episcopalian. My dad was raised as an American Baptist in Springfield, Massachusetts. He was a business major at Ohio State University. He became an insurance executive and was full of faith and integrity. I talked psychology with my mother, and sat at my father's side on many Thursday evenings while he read to my sister and me from Charles Foster's *Story of the Bible* (6) I learned to pray from my father and to think critically from my mother. For the sake of unity, my father became an Episcopalian but retained some pretty fundamental ways of looking at theology and particularly the Bible. I was beginning to wrestle with how to integrate the knowledge of the world wrought from reason and the stories that depicted a biblical faith. In retrospect, it was the best of worlds, faith and reason on the scaffold, each demanding its rightful place in the scheme of things.

I was fortunate to attend Berkeley Divinity School at Yale after graduating from The College of William and Mary in Virginia. Professors at Berkeley challenged many of the literal conceptions I held about the Bible as well as some of the orthodox theology with which I had grown up. After three years of intensive study, I began to appreciate that the Bible, among other things, was a profound book of spiritual evolution. In addition to studying it devotionally, I could also examine the Bible with reason and historical perspective. Like a thunderbolt, it struck me that the writing and reading of the Bible was a journey in which human beings and the divine God seek, reach and discover one another.

Over my lifetime, I have greatly valued what I have learned from science and what faith has allowed me to experience. At this point, to have one without the other would diminish my understanding of the world. Science and faith are not the same, but feed each other, complement each other, challenge each other, and expand each other. From each, similar and different questions arise. The answers, as well as we can discern now, are often not mutually exclusive. It is in the areas of overlap, where a degree of tension may reside, that we discover how each discipline has something to offer the other. We search with eager expectation for understanding and growth. For now, theologically based faith and ever-changing science still have much to discover because knowledge and faith move forward and the journey is open-ended. Tolerance and an appreciation of the mysteries of faith and science are required for the journey.

The fruits of an appreciation of science, particularly what I gleaned from astronomy, expanded my understanding of God, revolutionized my comprehension of the Bible, and enriched my capacity to cope with and enjoy the world about me. In our current culture, the polarization of these two fields of knowledge is distressing and is a principal motivation for writing a book that attempts to integrate the two noble disciplines.

The eighteenth and nineteenth centuries were replete with biblical scholarship, notably from German scholars, whose use of rational literary tools in biblical exploration came to be known as historical criticism. Their scholarship exposed the Bible as a living, not static, creation—the work of many authors who struggled to discern the will of God.

Only slowly did I appreciate and then embrace the idea that the writers' vision of God expanded and evolved from the Bible's inception to its conclusion. For example, early on, God was viewed as narrowly focused on the Hebrews. As the Hebrew writers accumulated a body of spiritual thought over the centuries and matured as a people, they embraced their God as the God of all humanity, and beyond that the God of the Universe. From the eighth to the fifth centuries B.C. the Prophets introduced the concept that the God they worshipped was more interested in justice and mercy than in the harsher ideas inherited from older cultures. Later, I will elucidate other evolutionary aspects of the biblical text.

In my own journey, my mentors allowed for grace, breadth and flexibility to breathe fresh air into the Bible. They also allowed science a full range of what it does best, to explore and discover, to make hypotheses and test them and to venture a little closer to material truths about the world and beyond. Jesus' own words have strengthened the union of science and faith in my mind. Mark's Gospel relates that when Jesus was challenged by one of the scribes, he drew upon the Hebraic Shema for his answer: "Hear O Israel, the Lord is One and…." In so doing, he reminded his followers to love the Lord not only with their hearts and souls but also with their minds. "You shall love the Lord your God with all your heart and with all your soul and with all your mind and with all your strength." (7) This last admonition is germane. God invites us to love Him with our mind as well as our heart and soul. We should not leave our brains at the door when entering the house of faith. Love of God does not mean sacrificing one's mind at the altar of faith. In fact it means just the opposite; if one loves God, that love will be enriched if one's mind enters the relationship. Mind and heart sing a bright new song when in harmony.

Now, allow me to tell you about my first crisis of faith. It was most acute when I was a sophomore in college. Due to an unusual circumstance, I had been corresponding with the Bishop of Puerto Rico, Irvine Swift, (8) for several years. By a quirk of fate or divine intervention, the Bishop was staying at our home in Longmeadow, Massachusetts, in 1951, having just returned from a missionary tour in China. I was eleven years old. The Episcopal minister's home was quarantined at that time because his son had polio. So he was asked to stay with us. Soon after he arrived, I announced to the Bishop, somewhat precociously and perhaps with excusable youthful arrogance, that I was going to be a minister one day. Little did I know what a good recruiter he was.

For the first few years following that visit, the Bishop kept in touch sparingly, but when I was in high school, his letters came more regularly and deepened in theological content. I rarely saw him. Puerto Rico is a considerable distance from Massachusetts so his long hand-written letters and my shorter responses sufficed. I became comfortable asking him about almost any aspect of life. He persuaded me that God was more interested in a loving relationship with plentiful forgiveness than the harsher notion of God that I had absorbed earlier. Over the course of fifty years we corresponded some 350 times.

In college, after initially immersing myself in philosophy, and dipping my toes into the waters of psychology, I settled on a major in the Department of Anthropology and Sociology. All of my professors, to the best of my knowledge, were agnostics or atheists. Most were in the "science of the uncommitted." Their disciplines derived power from the refusal to commit or believe in any one specific concept or idea. There was a certain intellectual comfort in this posture. I came to appreciate the discipline of thinking, and the value of doubt.

My faith was at a crossroads. Faith seemed to be entirely subjective; other religions seemed to have their own validity. What I was learning about science and world cultures caused me to doubt my rather orthodox, though not very well formed, view of the Bible. I was floundering in my own lack of direction, and my sense of purpose was diminished. So once again I wrote to my Bishop, confiding my desperate confusion.

He wrote back with a piece of advice that turned out to be the strongest and best piece of advice I have ever received. It seemed preposterous at the beginning, but I want to share it because it turned upside-down my notion about how faith was achieved. He taught me that the things of science are achieved in one way and the things of faith are achieved in another. Though the two seem antithetical, they are not; they are complementary, a part of the complexity and mystery of life, and both need exposure.

I suppose whenever anyone offers advice, it is taken, whether it is good or bad, only if one is ready to hear it. Often advice falls on deaf ears, but I was more than ready to listen. After all, I had solicited it. Ironically, at the time, I still wanted to be a minister. My brain had been fixed on this for years, but in some ways my goal was laughable. A minister who doesn't believe in God does not quite cut the mustard. And it was getting somewhat close to the time when I had to apply to seminary or find another field of endeavor. And so with some sense of urgency, I wrote to the Bishop.

As was his custom, he answered immediately. First, he asked me a simple yet bold and personal question: "Have you ever told the Lord that you loved Him?" I had not. He suggested that I begin by saying the simple words, "Dear Lord, I love you. Thank you for loving me." I cordially wrote back to the Bishop that this would be impossible as things stood. I did not love God, and I wasn't even sure I believed in Him. The hypocrisy of saying these words was intolerable. I would choke on them. The Bishop replied, "I want you to trust me. Say the words and let God handle the hypocrisy. Say them often and just see what happens."

Because of my desperation, I resolved to follow his advice for three months. My first hundred times uttering this prayer were like eating cooked broccoli gone bad; the smell and taste of the words were awful. I thought God, if there really were a God, would strike me dead for the hypocrisy of it. "Dear Lord, I love you. Thank you for loving me,"— such simple yet dangerous words. I choked on the prayer again and again. I loathed this drill. It made me angry, and it seemed to crush my spirit even more. I almost gave up, but I had promised myself a time period. I had a long way to go, so I kept at it. Slowly the pain eased. Sometimes my prayer was just by rote, sometimes I merely waved off the hypocrisy, and sometimes the prayer was uttered with great sincerity.

When the three months were up, I was free of the promise. It was time to give up the prayer. My reaction surprised me. I found I could not give it up. I began to say it on my own. I had begun to feel the relationship; it owned me.

Was it just the repetition? I have come not to think so. I surrendered, to be sure, but in the surrendering, I regained my bearings and my sense of purpose. I discovered that an academic relationship with God was not enough; it was valuable but it was more sociological than spiritual. What I discovered was a God who would not let me go even as I would not let Him go. It was somewhat like the story of Jacob wrestling with God's angel. As with fledging love or with any new relationship, the love itself would be tested. Growth would not come without some pain. The relationship between God and me would have to move from bedside prayers to real works of justice and mercy. Occasionally, an act of courage would be required that would reflect my level of commitment. Nevertheless, at age twenty, those simple words embraced my heart, and I couldn't say them often enough. "Dear Lord, I love you. Thank you for loving me."

I was introduced to this prayer more than four decades ago, and I have learned many since then. I have prayed aloud with many words, and I have prayed in silence. I have sometimes abandoned this prayer, but missing the relationship that I had with God through this prayer, I would come back to it. There is no doubt in my mind that this simple prayer given to me long ago is the most important prayer I ever uttered. More than any other, it nurtured the love of God in me. I cherish it to this day. Ironically, when I have felt secure in the love of my relationship with God, I have found that my conflicted feelings about the relationship between faith and science were not overwhelming. Some issues still presented a challenge but they no longer unduly threatened my faith. Faith and doubt began to exist with a certain harmony, and I was free to explore the riches of science and faith again.

IN SCIENCE, SEEING IS BELIEVING
IN FAITH, BELIEVING IS SEEING

If you have faith the size of a mustard seed,
you will say to this mountain,
'Move from here to there,' and it will move.

—Matthew 17:20

God... makes things make themselves. (9)

—Paleontologist and theologian Teilhard de Chardin

At the 2011 U.S. Open Tennis Championships, nineteen year old Melanie Odin and eighteen year old Jack Sock won the Mixed Doubles championship trophy. But it was at the 2009 United States Open when Melanie first burst onto the big stage. The relatively small seventeen-year-old was unknown outside intimate tennis circles and barely made it into the tournament's main draw. On her way to the semi-finals, Melanie upset three well-known stars, all highly seeded Russians. Two were former world number one players. Printed on Melanie's sneakers was the word BELIEVE in bold print. (10)

In the same month, it was reported that young golfing star Michelle Wie hired a new coach. He was not hired to change her strokes. His main goal was to make her believe in herself. Believing makes a difference.

The same is true of religious faith. The criteria for believing for people of faith are quite different than the criteria that science demands. Belief is not based on empirical observation. When I use the word believe, I will use it in the sense that one makes a leap of faith. It is interchangeable with the word trust. In certain matters of the human heart, and in many substantial matters of religious faith, belief or trust precedes the evidence. Yet, for those who take a leap into faith, whether it is a parent's trust of an adolescent behind the wheel of a car or the trust that partners have in a relationship, the perspective changes on both sides of the relationship. The gift of trust changes the attitude of the receiver of trust as well as the giver of that trust. And, in this change or affirmation, the world looks different for those who operate from belief or trust or faith. In matters of spiritual faith, believing enables one to see. Hence believing is seeing.

The Gospel of Mark relates a poignant scene in which the father of a boy inflicted with seizures brings his son into the presence of Jesus. He humbly implores Jesus: "If you are able to do anything, have pity on us and help us." Jesus responds, a little indignantly, "If you are able! All things can be done for the one who believes." Immediately, the father of the child cries out, "I believe; help my unbelief!" (11)

Science lifts up another perspective. Simply put, seeing is believing. It is well understood that to believe that something is true, a hypothesis must be put forward; it must be tested again and again, and its success must be repeatable at a later time by anyone who does the proper testing. Belief and time-tested facts go hand in hand.

If one plans to send a manned rocket into space, one must understand the laws of physics and follow them with prescribed accuracy. A rocket must be tested again and again, and decision makers must see plenty of positive results before risking space flight. At the same time, one may safely surmise that at the moment of liftoff, astronauts and their families pray without taking into account the empirical evidences of prayer. Astronaut Tom Jones, in his book *Sky Walking*, often attested to this. He notes that a priest was sometimes present to pray with them before launch. (12) Science and faith often merge at critical moments.

In the world of faith, few things are scientifically provable. Evidence of the Spirit's presence is not exactly repeatable for others to see. It is related to others through anecdotes and witnesses' stories, but that doesn't mean it is less real than the empirical demonstrations of science. It just means science and faith look at and experience the world in different ways. Most of the trouble between scientists and theologians occurs when one "side" tries to usurp the other's legitimate space. I suspect that many scientists and theologians are mutually guilty in this regard.

However, science and faith do appropriately edge into each other's boundaries, yielding some gray areas to mutual accessibility. Science is not 100 percent about seeing in order to believe, and faith is not 100 percent about believing in order to see. A scientist is also a human being who hopes and dreams and believes without being able to prove some of the most important aspects of living day-to-day. For example, an understanding that knowledge is good generally motivates scientists. They do not often make a sociological or even a philosophical argument for that tenet. Some instinct or natural human trait exists in their inner core that persuades them that what they discover will be good or helpful or useful, if not in the short run, certainly in the long run. Scientists proceed each day without being able to prove this fundamental motivating factor.

One does not usually become a scientist without caring about the world and trying to improve it. Albert Einstein deliberated about whether or not the universe was static or expanding and whether the universe reflected a sense of purpose. He worried that a universe that expanded or drifted off and dissipated into nothingness made no sense. Likewise, scientists want to live in a purposeful universe as much as anyone else, so most go about their day-to-day lives carrying a dollop of faith. As with anyone else, when they lose that faith and sense of purpose, their work and their raison d'être suffer.

The world of human relationships merges experience and belief together all the time. If a parent believes in a child's goodness or abilities even though there is little observable evidence for either trait, the belief can eventually enable the child to mature and succeed. Social scientists tell us that a priori belief in children is necessary to give them the confidence to live well. Belief and success are regularly connected.

One of the themes in this book is that the world needs the perspective of people who are willing to trust the human experience and to see in it intimations of the Divine. Belief or faith or trust can determine one's perception of the world. A religious perspective or a faith perspective is grounded in this understanding. An old Eastern tale illustrates this idea.

A weary traveler was looking for a place to rest. He came to a crossroads and encountered a wise old man. The traveler inquired about a town he saw in the distance. "What is the town like?"

he asked. "What sort of people live in that town?" The old sage responded, "What are the people like in the town from which you have come?" The traveler answered, "Oh that town was filthy and the people were vicious. Besides, the town is full of thieves." The sage spoke again. "Well, don't go to the town you see in the distance, avoid that town; it is filled with the same kind of people. You'll find no peace there." The traveler thanked him and headed off in another direction.

Not long after, another traveler approached the same crossroads and met the wise old man and asked the same question. "From what kind of town do you come?" asked the sage. The second traveler responded, "Oh, the town I came from has streets lined with flowers, and the people are warm and generous; nearly everyone greets you with a smile." "Well," the wise man responded, "you should definitely go to that town you see in the distance; it is just like the one from which you came; you will find the same kind of people there."

How we experience and engage with the world in large measure reflects our own perceptions, disposition and attitudes. People who believe that human beings have goodness in them will find goodness in other human beings. By contrast, people who are inherently suspicious of other humans find much to be suspicious about.

The same is true of those who believe in God. The "believer" will find ample evidence of God in day-to-day life. The "nonbeliever" will find ample evidence that God is absent or nonexistent. Maybe no more vivid example of this occurred than in 1961 when Yuri Gagarin, Soviet astronaut, the first man in space, rather smugly declared that when he was in outer space he did not see God. After Alan Shepard's first American space flight, later that year, a chorus of our early astronauts declared just as brazenly that God's handiwork was obvious from outer space. The two sides answered with the bias of their own perspective.

Science and faith need to be separate in many respects. Each must go about its way of understanding the world according to the tenets that make its discipline function and succeed. Each must know its own boundaries. Also, each needs to know what can be shared with the other and where each can benefit from the other. I believe some flexibility is needed in the areas of science and faith. In pure science, a healthy skepticism is necessary to avoid jumping to hasty and incorrect conclusions. In one's attitude toward life itself, belief often precedes success. The two are not mutually exclusive.

Two examples may serve this thesis. An archeologist makes a discovery as in the 1947 Dead Sea Scrolls. Science plays a key role in the process of discovering and discernment. An inquisitive and open mind is necessary to extract the data and interpret the findings. What attitude may a person of faith bring to such discoveries? Faith can be open to the challenges without being threatened, for faith does not rest mainly on material evidence, but rather stems from an inner trust that God is in all things. With that open mind, a person of faith may alter his understanding of a piece of history and feel enriched in the process. Even when some supposed religious facts are challenged, faith is not at risk, for faith is not dependent upon supporting evidence. Faith is a decision to trust God no matter what.

Likewise, scientists may come into the realm of faith without shedding their skeptical mindsets. They may listen to a fine production of Handel's Messiah and be every bit as swept away by the spiritual beauty as a traditional believer. They may pray as earnestly as anyone else, recognizing the mystery of that which they do not know through the methods of science. Science and faith are the products of human experience. I contend that the tree of knowledge that supports both disciplines must be flexible enough to bear the fruits of each.

A BREATH OF FRESH AIR BLOWS THROUGH THE BIBLE

I will lift up mine eyes unto the hills...my help comes from the Lord,The maker of heaven and earth.

—Psalm 121:1

The Bible is a book about how one goes to Heaven—not how Heaven goes. (13)

—Cesare Cardinal Baronio

The Bible may be studied as an evolutionary spiritual document for a maturing world. It may also be appreciated as a historical dialogue between the early Hebrew/Christian people and God. If the Bible can be regarded as a work in progress—the journey of a "chosen people" with their God, a story with a spiritual progression—readers everywhere could engage the Bible anew. A breath of fresh air would, indeed, impact all houses of worship.

People who think the Bible is God's infallible literary blueprint may initially be disappointed at the suggestion that the Bible is richer and more meaningful if it is allowed to "breathe." The breath of the Spirit, ruah in Hebrew, is characterized by a wind. A wind blows and moves stagnant things; the old doesn't stay the same, a people imprisoned can be set free. Likewise, the Bible is an instrument that plays better when the Spirit is allowed to live and move and breathe. For example, if the Bible's Creation story, particularly about Adam and Eve, were accepted as a myth, we would not be so stuck on the age of Earth or when the first humans walked on earth; we could concentrate on the story within the story, which is the relationship humans have with God and the obedience necessary to healthy living. If the six days of creation can be accepted as allegory, more people would be open to the wonders of science that describe an entirely different timeline and scope of creation. Many others would entertain reading the Bible rather than regarding it as outmoded, primitive material. Breathing fresh Biblical air means being open to the human journey: open to science, music, letting go of fixed prejudices, seeking new knowledge, and addressing modern social problems as illustrated in a future chapter.

Myth is sometimes a confusing word because it is used in difference senses. When I use the term, I do not mean that the story it tells is not true. I use myth in the sense that the story it tells, while not factually accurate, reveals more truth and is of more value than facts can disclose. A good myth touches the heart and speaks to the spirit within humanity. For example, when a serpent speaks we feel the eerie power of a dark force. We do not have to know what dialect he chooses. Grand myths reveal truths that transcend our ordinary rational understandings. Myths can speak of God in ways that facts and ordinary history fail to convey.

It may be easier for some to assume that the Bible is God's document alone, that it is a perfect book, in which each line is the way God would write it if the Almighty had pen in hand. On one level, it is, perhaps, intellectually convenient to believe that God dictated the Bible from A to Z or inspired the writers to write directly from His hand. One doesn't really have to use one's mind after that. One doesn't have to explore the discoveries of science or to experience the riches of many other fields of learning. More importantly, one doesn't have to engage the writing with one's mind and one's faith.

Because of literalists, the Bible suffocates under its own weight. Those that insist that every jot and tittle is from God's inerrant pen must expend an inordinate amount of effort defending their point of view against the onslaught of scientific evidence. At risk, so the literalist thinks, is his or her faith. In contrast, with a deepened and perhaps more mature faith, one can relax a bit and experience the beauty of the journey that human beings take with God in the mutual, though not flawless, writing of the testaments.

Literalists also believe that revelation begins and ends with the Bible. Allowing the breath of the spirit to flow through it allows and permits readers in all ages to participate. Revelation is not static; it continues, and we are able, in our generation, to participate in it. Believing that God revealed all that is important between 1500 B.C. and 100 A.D. burdens the Bible with too much weight.

The most startling example, in my experience, of forcing a literal understanding of the Bible's words upon a modern scientifically oriented world is captured in the following anecdote.

After twenty-five years of pastoral ministry, I spent a sabbatical year in the Florida Keys, where I taught tennis at a small club and attended an Episcopal church on Sundays. I was asked to teach the teenage church school class, and I gladly accepted. I knew that most of the class attended the weekday Christian Bible School, but I was, nonetheless, surprised by the students' comments when the subject turned to Creation. I attempted to reveal what I thought was important about God as Creator, while expressing this in the context of a universe that had existed for billions of years. The students were skeptical. Most of them had been taught that the universe was but a few thousand years old.

In an attempt to lead the class in a more open direction, I shared an experience that my wife, Jeanie, and I had during the previous summer on our vacation. We had flown to the Grand Canyon and were staying at the scenic El Tovar, a rustic log inn on the South Rim. On the day after our arrival, we began our trek down to the Colorado River via Bright Angel Trail.

As we descended into the canyon, we marveled that we were also descending millions of years in geologic time. Every few hundred feet down, the geologic timelines went back between one hundred and two hundred million years or more. Along the way a ranger occasionally pointed out the evidence of fossils. Just a few thousand feet below the rim, in an area that had been on the earth's surface a billion and a half years earlier, the visible fossil evidence came to an end. Either the conditions to make fossils were not present earlier or there was no prior visible life. It was clear, at least in that portion of the earth where fossils existed, that our planet was old, and recognizable life had existed for more than a billion years. The testimony remains in the rocks. Paleontologists have, of course, radiocarbon dated many fossils to over a hundred thousand years. Geo physical analysis can date volcanic ash to millions years and genetic measures dates that far back as well. (14) Human history is old too. World famous paleoanthropologist Richard Leakey writes…"beginning some

2.5 million years ago...Homo Erectus represents a major turning point in human history. From the vantage point of today, it speaks of leaving an essentially apelike past and embarking on a distinctly humanlike future." (15)

That evening we enjoyed a lecture from one of the rangers about the history of that extraordinary piece of land eroded by time and the river. We slept that night at Phantom Ranch, some six thousand feet below the canyon's rim. The next day, tired but still exhilarated, we climbed out. When we reached the top, we felt as gleeful as marathoners finishing a race. Then, exhausted, we collapsed on our beds. We had taken a marvelous journey into the crust of the Earth, and it was awesome.

When I related that experience to my Christian Bible School students, their reaction amazed me. Most were skeptical. They had been taught that the world was about six thousand years old. They were, perhaps, referring to the well-known calculations of the Archbishop of Armagh, James Ussher. (16) He was the Irish prelate who had, in the mid-seventeenth century, added up all the ages of the biblical family, such as Adam at 930 years, Noah at 950, Abraham at 175, and so on. He came to the conclusion that God created the world in 4004 B.C. (17) He was not alone. Other great Biblicists of the era added up the chronologies from Adam to Jesus and came up with similar dates. For several generations following, many churches accepted that timescale.

When I confronted my students with the "rock's testimony" in the walls of the Grand Canyon, they were initially resistant. They said they had been taught that God placed the fossils in the ground just as they were, and that what I had seen in the Grand Canyon was God's handiwork from about 4000 B.C. Rather flabbergasted, I asked this simple question: "Why would God do a thing like that? Why would God take the time and effort to make fossils from scratch and stick them in the ground at various levels?" The class did not know immediately but said they would come back with the answer. The next Sunday we met again and they were ready. The collective answer was that God placed the fossils into the ground in order to humble human beings. We are too proud and God sends us to our knees. He humbles us through this godly deception. They continued with words to this effect: We humans are led astray by our intellectual arrogance; only faith in the inerrancy of the Bible could save us. The class identified this way of thinking as the "Mature Earth Theory." (18)

I was not only their Sunday School teacher and youth advisor but was friends with their parents who were persuasive in asking me to be the interim minister during the illness of the rector, the Rev. Joseph McDowell. In the course of our pastoral relationship, I believe I was able to show them that the price was too high to maintain a belief that the earth was so young. They would have to deny the sciences of Geology that affirmed the oldness of the earth as well as Paleontology and the ancient dinosaurs, Astronomy and Astro-physics, Archeology, and spectacular ancient finds. Why, even the late Woolly Mammoths were older than the earth's origin that they were suggesting. More importantly, I hope I was able to show them a better way, a God who was truly grand, whose creation was vast and whose time was eternal.

In regard to the ages of the early men of Genesis, the ages 930 for Adam, 950 for Noah, 969 for Methuselah and even 175 years for Abraham lack credibility. One school of Biblical thought contended that since man sinned, the ages of life kept getting smaller. The scientific community would argue that early man lived fewer years than we do but that for several thousand years the age has been pretty stable. The Psalmist (90:10) counts three scores years and seven to be about average. Interesting that in America the census shows ages creeping up again. By the old calculation, we must be sinning less.

But the Sunday School example from 1994 made clear to me that the unshakable literalist will go to extraordinary lengths to protect his point of view. To be fair, what the literalist believes he is doing is protecting the Faith, and he would die for that cause. What I would like to express to him is that there is a better way that incorporates other truths that come from science. There does not have to be a polarity between the points of view. Faith is not at risk. I can love God as much as the fundamentalist can with his point of view.

Faith is choosing to believe and to love God without proof. Faith involves a leap. Belief does not suffer with such an approach. Seekers of the "Truth" need not fear an examination. The wonderful news is that the honesty is rewarded. A richer faith beckons. Loving God and loving Jesus is not dependent upon Biblical literacy, just as loving a person is not dependent upon their age or even their looks. Loving God comes from the soul and is practiced with one's will. There is plenty of love and insight in the Bible, but it is not a scientific, chronological proof text. It is a living document, which means it goes beyond rigid interpretations. There is something new to be discovered with each reading.

Loving God with one's heart and soul usually, and perhaps necessarily, precedes loving God with one's mind. But, once the love is real, appreciating the gifts of the mind and reaching out with the best tools of one's intellect enhances and deepens the love. The intellect teaches the vastness and depth of God's world. God becomes big enough and grand enough to encompass all of it.

Somewhat ironically, after my sabbatical in the Keys, I became chaplain and teacher at Salisbury School in Connecticut where I had been a student thirty-five years previously. I taught seniors theology and astronomy. I found a very different take on this question. Almost no one in eleven years of teaching there believed the earth was young. Most students readily accepted an ancient earth and an even more ancient universe. Half were believing Catholics, and the Protestants were mostly still trying to figure things out. The Jewish boys had no trouble integrating their Biblical faith with science. My challenges as a teacher were mostly to bring faith into their lives; the science was easy. My experience was very nearly the opposite challenge from the one in the Keys. I learned first hand what great differences are in our religious education depending upon where our homes are located and the traditions that are dominant in those locations.

With this point of view guiding us, one embarks on a biblical journey, a varied path, where roads divide and then come back to a center, and something more profound emerges. I believe that the Bible ought to invite one to participate in ongoing revelations during one's lifetime and to encourage a theological walk every bit as important as the walk the prophets took thousands of years ago. If one believes that revelation has not ended, then one can seek new solutions to old problems or be sensitive to challenges that were not even recognized in the past. Today most would agree that the institution of slavery is anathema to the walk God's desires for us. But it was not such a keen issue in Biblical times. It was a different time, a different history, and the question of slavery had not yet risen to paramount importance as an ethical question. It would take a long time before it did. Each generation can look for the issues which rise to the surface in their lifetimes. In my church, I have seen many such issues, none bigger than raising women to the level of priests more than thirty years ago. Our church was participating in the ongoing revelations of the spirit.

I suppose the greatest fear for literalists is that if they stop believing that the Bible is God's word without any qualifications, then where do they find their faith? If they let themselves believe that the Bible was not dictated to men by God's almighty pen, there is no basis for faith.

A better interpretation is that faith is birthed by the Spirit that God has given to us. The answer comes from the human conscience. The answer comes from experience, from life's tests, from the wonderfully written but imperfect Scriptures, with the testimony of a "chosen people" for over a thousand years. Basically, those people chose God repeatedly during their very earthy history. With all their joys and sorrows intermingled, they wrought a faith and a way of life, an ethic to live by and a Lord to worship.

Their faith evolved through the living of life and devotion to the belief—without benefit of a written Bible for most of that time—that God was real, that God's Spirit dwelt in His creation, and that His Kingdom dawns on Earth and never ends.

What would it mean if more people espoused this more open-minded biblical perspective? I believe the Bible would be less threatening to more people. I believe more people would read it, study it and appreciate it. If people of faith and others with curious minds could adopt an understanding of the Bible that was open to interpretation, it would be a more engaging book. People would be free to think, participate, engage, wrestle and wonder, and the Bible would live more vibrantly than it does now.

IN CELEBRATION OF BIBLICAL EVOLUTION

Come now and look upon the works of the Lord,
what awesome things he has done on the earth.

—Psalm 46:9

The whole sense of evolution…is the ascent toward personality, we are living in a personalistic universe. (19)

—Tielhard de Chardin

What difference does it make to ordinary human beings in the twenty-first century if the Bible is read with the understanding that its texts and tenets evolved through the centuries or if it is considered as a document without human flaw? This chapter will attempt to shed light on this question.

I will move quickly past the scientific idea of Creation and the literal biblical Creation, except to make this general statement: The modern theories of science cogently describe how the universe expanded and evolved after the first few moments of Creation. We have come to appreciate the eons involved in the process, as well as something of the grandeur of it all. Many religiously oriented people give plentiful thanks for what scientists have discovered and constructed over the past several hundred years. Others are suspicious of or threatened by the incomprehensible vastness of space and length of time of the universe's existence. I join with many who rejoice in the discoveries of scientists and try to absorb and appreciate their theories. Sometimes this means tweaking or adjusting biblical and theological points of view that I may have held in the past.

With the Bible's opening words, "In the beginning, God created the heavens and the earth…," the writer is making a faith statement. (20) He offers no proof and gives no data to support his statement. It is germane to note as evidence of an evolving Scripture that the opening chapter of Genesis was written much later than many other sections of the same Scriptures. The New Interpreters Bible states this: "Many scholars consider the opening two chapters of Genesis as two creation stories, assigning 1:1-2:4a to the Priestly writer and 2:4b-25 to the Yahwist." (21) The Yahwist, the earliest of known Hebrew scribes wrote in the eighth century B.C and the Hebrew Priestly writers wrote in the fifth century B.C. The first chapter was written well after the Hebrew people had experienced many of the joys and sorrows that comprise the Hebrew Scriptures today. Therefore, the opening passage has the weight of a people steeped in generations of spiritual traditions.

However, the story of Creation in Genesis is not a scientific document. There is no attempt to prove either the existence of God or the when and how of Creation. The writer enthusiastically and unabashedly declares that God is the Creator of Heaven and Earth, and that Creation in all its parts is "good." When this early biblical editor writes that God created the "world" as we know it "in six days," he is creating the mood of faith through poetry and analogy. He is not wedded to a chronology, certainly not a twenty-four-hour day. If we want to know how long Creation took, we would do well to turn to the work of science and appreciate the wonder and magnitude of some thirteen or more billion years of time, measured by our earthly standards.

But when we want to know why we were created, we may turn to the Bible. There we will discover the experience of a people of faith whose lives uncover the meaning of human relationship as well as the relationship that is possible with God. When people want to know how to communicate with God, the two Testaments can guide them. The Hebrew Scriptures offers the centuries-long journey of the Hebrew people; the Christian Testament relates the recollections of the life of Jesus and the life of the fledgling Christian community. Together, those Testaments expose the breadth and depth of human experience and are replete with prayers and pastoral advice for every imaginable need or occasion. For example, when people grieve over the way they have lived, they can turn to the Bible for its understanding of repentance and forgiveness. When nations are contemplating war, the Bible offers enormous insight and may remind people of what they have learned about God and His purposes.

In support of an evolving Scripture, let us consider four examples of just how biblical writers, particularly in the Hebrew Scriptures, amended the thought of previous generations and reached for higher ethical or theological positions.

The first example is simple, yet important, because it constantly appears in popular culture as a divisive issue between scientists and religious people. It concerns the first and second chapter of Genesis, the Creation and a story of "the first human beings." To understand how this story came to be, one must first recognize how the Scriptures were put together. The Scriptures were written over many centuries. Culture evolved and with it religion and a more complete understanding of faith. The writing of Genesis, as well as much of the Bible, is not a straight-line ascent of progress, but read with care, the spiritual growth of the Scriptures becomes evident.

One of the most misunderstood yet important discoveries of biblical scholarship is that later writers did not automatically discard and delete the writings of earlier authors. For example, a fifth century B.C. writer having a scroll from the eight century B.C. often chose to leave the earlier writing nearly intact but write his own take on the subject at hand before or after or sometimes woven into the earlier writings. Thus, today, within the Hebrew Scriptures, we may read the older and newer writings together. Only careful analysis can differentiate one from the other. We can appreciate from where the Hebrew writers started and how their writings grew and developed over time.

The amending process is evident at the very beginning, with the first and second chapters of Genesis. The first chapter was written several centuries after the second chapter. Scholars have determined that the earlier writer, in the second chapter, known as the Yahwist, was a storyteller who beautifully personified the relationship between God and humanity. He told his "Creation" story as a myth to reflect an introductory engagement between God and the first humans with a God-consciousness. Strong myths often reveal truths that that have more lasting impact than factual stories have. Chapter two remains an elegant story from which much about God and His Creation can be learned if one looks in the right direction. Chapters two and three relate the story of the first man and woman in the Garden of Eden. In part of the myth, woman is created from that infamous rib of the man, to be his helpmate. The emerging conscience of men and women is introduced and many questions are raised. What does it mean to obey and disobey God? What are the consequences of their behavior? What opportunities are available to humans after serious breaches of trust? The Yahwist invites an intimate picture of the meeting of God and humanity. He portrays God as personal and within reach of ordinary human beings; God walks with man in the Garden and enjoins him in dialogue.

A few centuries later, as the body of the Hebrew Scriptures developed through the work of scribes, priestly editors were heavily involved in putting the Scriptures together in an orderly way. They were concerned, as well, with temple matters, the rituals used in worship, and in helping the community find a standard of worship for all the Hebrew people. In the process, they continued the development of the early myths of the first humans. They were inspired to place a chapter before the Garden story, to leave the early tale basically intact, but to write a much more sophisticated version of Creation. As a result, they made some wonderful leaps forward.

Their understanding of how life evolved on this planet was quite amazing, considering that they did not have our modern scientific tools. The author, with a sense of poetry and depth, begins: "and darkness covered the face of the deep…and God said, 'let there be light'." (22) Then the biblical scribe tells us that there was darkness and then light, there was water from above and beneath, there was dry land and then vegetation: "And God said, let the waters bring forth swarms of living creatures (23) The progression continued: There were fish and birds, then creeping things and wild animals and finally humanity. In this later version by a priestly editor, man and woman were created together. There was no more rib of Adam, prescient, perhaps, of efforts toward equality between men and women. It is difficult to imagine a greater leap in the emancipation of women than that single amendment to the biblical text—from the thought that Eve was created from Adam's rib to be a helpmate for him to the revelation that man and woman were created together.

Beyond the early description of how life on the planet took shape, this priestly editor made some observations that are equally important, yet not often mentioned. At the end of each day of Creation, according to the author, God surveyed what He had created and declared that it was "good." There is nothing here that would indicate that Creation was tainted. God was pleased with His Creation. His godly assessment of the total process according to the author was that, "Indeed it was very good." (24)

The concept that the Creation was flawed and that humans were created in sin is not found in the biblical text. On the contrary, the opening chapter of Genesis celebrates again and again that Creation and the creative process is "good." Dominican scholar Matthew Fox makes this point splendidly. "The Universe itself, blessed and graced, is the proper starting point for spirituality. Original blessing is prior to any sin, original or less than original." (25)

For the second example of an evolving Scripture, we now plunge deeper into the text. Most biblical scholars agree that the Ten Commandments were given and written down long before the many laws that one finds in the Book of Leviticus. The Ten Commandments were a kind of Emancipation Proclamation when they were first given to the Hebrew people. For example, let us look at the fourth commandment. In Egypt, as well as in the early days of living in "the promised land," the Hebrews were used to having to work for slave masters or overseers seven days a week. The fourth commandment states, "Remember the Sabbath Day, and keep it holy" (26) (later scribes added: "In it you shall not do any work"). This addendum was a clarion call for emancipation. Because various overlords controlled them, it was difficult for the Hebrews to make that commitment, but they took a revolutionary stand and often suffered the consequences. The ancient Hebrews believed that just as God "rested" on the seventh day, so should they.

I would note that many scholars believe, as well, that the Ten Commandments evolved from such documents as the Code of Ur-Nammu, the Sumerian King of Ur, in the twenty-second century B.C. and from the Code of Hammurabi, the Babylonian king and lawgiver from the eighteenth century B.C. (27) This borrowing and building on to past great works is similar to how Thomas Jefferson immersed himself in French writings about independence and then went on to compose our Declaration of Independence. What Jefferson learned from the French enabled him to write that momentous document for the fledgling American colonies. In the same way, the early codes of human conduct had influenced the early Israelites and the writing of the Ten Commandments.

In most societies, early simple laws or codes grow in complexity over time. If left unchecked, this growth may eventually stifle the human spirit. We amend laws as people find loopholes and, as a result, more laws are written. This happened with the Ten Commandments. The ten laws issued by Moses at Sinai seemed sufficient for a while, but by the sixth century B.C., more than six hundred laws appeared in the biblical Book of Leviticus. The Ten Commandments remained the overall guide.

By the time of Jesus, many of the laws surrounding and amending the commandments had become too legalistic and some thwarted the spirit engendered in the original commandments. Jesus addressed some of these issues. As an itinerant Jewish teacher, he was able to cut through them with the sharpest of spiritual scalpels. His spiritual genius was not to create new or better laws than already existed, but to pick from the very best of his own Hebrew Scriptures and re-create the spirit of the law. For example, in one teaching moment, he reduced all the commandments to two. He chose the first from Deuteronomy (28) and the second from Leviticus. (29) Together these are often called, "the Great Commandments." This is how they are combined in Mark's Gospel. "Hear O Israel: The Lord our God, the Lord is one; and you shall love the Lord your God with all your heart, and with all your soul, and with all your mind. This is the greatest and first commandment. And a second is like it: You shall love your neighbor as yourself. On these two commandments hang all the law and the prophets." (30)

The third example that demonstrates the way in which the Scriptures evolved comes from the ancient book of laws, the third book of the Torah or the Pentateuch. The law called lex taliones, the law of equal retaliation, replaced a much harsher one. It reads as follows in this Scripture: "Anyone who maims another shall suffer the same injury in return: fracture for fracture, eye for eye, tooth for tooth; the injury inflicted is the injury to be suffered." (31) We may note that if the Bible is viewed as a document in which no growth is allowed, then this ethic would remain the Biblical standard today.

It is often forgotten in our contemporary society that the law of equal retaliation was a great advance for justice in its own day. It represented emancipation from earlier, crueler laws that were uneven and weighted in favor of those who were more privileged. Lex taliones heralded a new code of fairness and applied to everyone, irrespective of one's station in life.

And yet, Jesus saw a better way. His ethic was a radical departure from this well-accepted rule of fairness. The new ethic is reflected in the Sermon on the Mount which the early Christian community thought captured the essence of Jesus' ethics:

> "You have heard that it was said, 'An eye for an eye and a tooth for a tooth.' But I say to you, do not resist an evildoer. But if anyone strikes you on the right cheek, turn the other also; and if anyone wants to sue you and take your coat, give your cloak as well; and if any one forces you to go one mile, go also the second mile." (32)

This sublime ethic is so heavenly that many think it is of no earthly use. It, indeed, is a lofty form of love that can produce such an ethic, yet this ethic is not unknown in nature. For instance, I was long ago impressed to read about how wolves behave in the midst of a fight, often in a fight over territory. When one wolf has had enough, he surrenders by offering his throat to the other. The other does not take it. He has won, and there is no further injury needed.

In human confrontations, we also can offer the other cheek if not our throat to another. Mercy is a supremely spiritual characteristic. There is some noble force within most of us that keeps us from hurting someone when he is most vulnerable. It shows strength of character to turn one's cheek, or our collective cheeks as a people, in order to transform a confrontation into a negotiation and, ultimately, a reconciliation. Obviously any critic can cite chapter and verse of those who have hurt innocent victims. But, in day-to-day living, turning the other cheek demonstrates human character at its best and can inspire similar conduct in others.

As we examine the evolution of ethics, sometimes we find that compassion and love were served better in the past than in the present day. Evolution is not a straight line forward. There are setbacks from time to time. Nevertheless, we can usually recognize goodness and progress when we see it, and this principal is applicable to the Scriptures as well. Sometimes the same ethical argument must be argued again and again. One may see in our own day, for example, how we tend to vacillate in our opinions and legal judgments of capital punishment or abortion.

In a similar vein, during most of human history it has been easier to hate one's enemies than to love them. The terms "enemy" and "hate" are practically inseparable. Each takes on meaning in relation to the other. It was acceptable, throughout much of the Scriptures, to despise one's enemies and pray that God would inflict them with the punishments they deserved. Many of the psalms call down God's wrath upon Israel's enemies; for example, Psalm 59:11: "My merciful God comes to meet me; God will let me look in triumph on my enemies. Slay them O God, lest my people forget; Send them reeling by your might and put them down, O Lord our shield." (33) This is certainly a natural expression of how many of us can feel when we believe we are attacked unfairly by either a person or a nation.

In the Christian Testament, it is clear that Jesus reaches for a higher ethic. Well versed in the culture into which he was born, he knew well the law of fair and equal retaliation. He offered a minority view, not unknown in the Hebrew Scriptures, but a point of view he brought to prominence: "You have heard that it was said, 'You shall love your neighbor and hate your enemy.' But I say to you, love your enemies and pray for those who persecute you…for if you love those who love you, what reward do you have?…And if you greet only your brothers and sisters, what more are you doing than others?" (34)

In contemporary society, we find it terribly difficult to give more than lip service to this ethic. When attacked, we feel that it is morally acceptable to strike back. We do not necessarily follow the ethic of the more evolved Scriptures, but often resort to more primitive ideologies. Imagine what it might mean to live by this higher ethic. It was a daring idea when Jesus proposed it; it remains a daring concept today. It just might be that the world could be healed of its greatest ills if we would risk acting according to these gracious ethics.

The fourth example of evolution within the Scriptures can be seen in the development of the concept of the messiah. The prophets of the Hebrew Scriptures set the moral and ethical tone of Israel. In so doing, whether railing against kings and princes or speaking to the conscience of Israel, they advanced the cause of justice and mercy. In this milieu, messianic ideas were born. This evidence of evolutionary progress is witnessed in the prophecies of First and Second Isaiah, each housed within the single book of Isaiah.

Preaching to Israel in the mid-eighth century B.C., Isaiah's ministry began when the king Uzziah died. First, Isaiah's ministry extended over the reign of four kings. In that ministry, he offered praises at inaugural celebrations for the new king. It is thought that at King Hezekiah's inauguration, Isaiah offered a description of the ideal king. Centuries later, in Christian settings, this was seen as a messianic foretelling: "For a child has been born for us, a son given to us; Authority rests upon his shoulders; and he is named Wonderful Counselor, Mighty God, Everlasting Father, Prince of Peace. His authority shall grow continually…." (35)

This early conception of a messiah held that the "anointed one of Israel" would be an earthly prince of great power, recognized for the magnificence of his work and his being. As the conception of a messiah developed in the ancient Hebrew culture, Isaiah's description of a righteous king became the standard for what was hoped for in the future.

A century and a half after First Isaiah, while many Jews were living in exile in Babylonia, their prophecies, hopes and dreams did not die. Jeremiah and Ezekiel were among those who helped to keep them alive. In that period, a rather obscure prophet arose. He is known to us today only as Second Isaiah, but he contributed one of the most advanced thoughts in the Scriptures, a new vision of what the messiah might be like. In Second Isaiah's vision, the messiah was not a mighty warrior nor was he a king robed and glorious in his power, but a humble man, a suffering servant, a man eventually broken and sacrificed for the people.

> "He had no form or majesty that we should look at him, nothing in his appearance that we should desire him. He was despised and rejected by others; a man of suffering and acquainted with infirmity…. Surely he has borne our infirmities and carried our diseases; yet we accounted him stricken, struck down by God, and afflicted. But he was

> wounded for our transgressions, crushed for our iniquities; upon him was the punishment that made us whole, and by his bruises we are healed." (36)

This is one of the more subtle evolutionary advances, for it trades power for vulnerable love. It is a concept we have not yet digested or fully accepted in the "real" world. We seem to save that concept for circumstances when the cost is not so great. We love to hear Second Isaiah preached from the pulpit or listen to it as in Handel's Messiah. We may even preach it to our kids as when we train them to walk away from a fight even if it costs them a moment of humiliation. However, when push comes to shove, as when we perceive a threat to the nation, we still often vote for physical power over vulnerable love. Soon, lest we perish, we must find a way to opt for the latter.

I have offered these few, but important, examples to illustrate that the Bible is a living, evolving document. Its concepts and codes of conduct have been amended within the Bible itself. Yet, it is wonderfully fortunate that the Bible's own early editors and redactors left intact much of the early writings, even as they amended and added on to the Scriptures. The merging of former writings with later writings allows us to enjoy the Bible's growth and evolution. If we understand how the Bible was put together, we can gain new insights, as active participants, and the text remains fresh and alive. The Spirit can yet teach us new things, and the attentive reader may be ready to hear God's revelation for this current age and beyond.

THE FAITH OF SOME GIANTS OF SCIENCE

Let us now sing the praises of famous men,
our ancestors in their generations. (37)

—Ecclesiasticus (Sirach)

Everyone who is seriously involved in the pursuit of science becomes convinced that a spirit is manifest in the laws of the Universe... a spirit vastly superior to that of man, and one in the face of which we with our modest powers must feel humble. (38)

—Albert Einstein

Some of my attraction to the potential integration of science and faith stems from the theological convictions of many scientists of note who espoused both disciplines. While many of them eschewed the limited understanding of science common to many churches, as well as some of the doctrine of those churches, they often supported the overall faith of church people and were deeply grounded in faith themselves. Because of their greater perception of the workings of the universe than most people, they were more in awe and glimpsed more dimensions of the Creator's majesty than most people.

At the same time, each of the scientists that I will herald as people of faith had slightly differing concepts of Almighty God's role in the great scheme of things, from the Creator's role in Creation itself to whether there was a divine ongoing presence in current events. Some had been slated early in life to be clergymen or came from backgrounds that were immersed in church life. Some called themselves deists; none would identify himself as an atheist.

I have chosen some of the icons of science: Nicolaus Copernicus, Galileo Galilei, Isaac Newton, astronomer Maria Mitchell, and Albert Einstein. They revolutionized the science of their day and changed the common perceptions of the world for ordinary mortals. I will illustrate briefly how they integrated faith into the body of their work.

I begin with Nicolaus Copernicus, who was born in Torun, Poland, in 1473. Though standing on the shoulders of some ancient astronomers, Copernicus is the one who convinced much of the world that the heliocentric system surpassed the geocentric Ptolemaic system, which had been articulated by Aristotle and accepted by most scientists for centuries. (39) His new theory, of course, overthrew the common sense of millions of ordinary observers. One could merely go outside and watch the sun move across the sky from east to west and do the same thing the next day. How counterintuitive it was to suggest and prove that it was the other way around – an outrageous idea that the mighty Earth revolved around the sun. Copernicus was not a professional astronomer; he was a doctor of canon law. He had studied medicine and mathematics; and was considered a Renaissance man. For forty years, he made his living at the cathedral at Frombork (Frauenberg). There was no contradiction for him between science and faith. Reordering the Catholic Church's understanding of the celestial positions of the Earth and the sun had no bearing on his belief in God. For Copernicus, the creative mind of God was even grander in the more simplified system he proposed. It meant that God was a more refined architect of the universe. Copernicus died in 1543. The Catholic Church did not seriously attack his theories while he lived.

The second icon of science and faith is Galileo, born in 1564, twenty years after Copernicus died. Like Copernicus, Galileo was a strong Roman Catholic. He lived during the papacy of a dozen popes and experienced, as might be expected, different relationships with them. Pope Paul V, in office for sixteen years, had promised Galileo protection for life. When he died, Gregory XV, a very frail pope who survived only two years in office, succeeded him. Pope Urban VIII succeeded him and served as pope for more than twenty years. He is the pope who held Galileo under house arrest. Urban had tolerated Galileo's views as long as the scientist did not press his position too strongly and until some of the Inquisition's powerful proponents leaned on the pope to be more active. (40)

Meanwhile, Galileo was putting his newly made refracting telescope to astronomical use and his findings confirmed what Copernicus had proposed. The timing wasn't great; the Roman Inquisition, emboldened in 1542 to deal with Protestantism, was heating up. Its proponents had now spread its zeal to deal with aberrant views even from Roman Catholics such as Galileo. The Catholic Church had not paid particularly close attention to

Copernicus' views, which were supported by Galileo, but now the Catholic Church denounced the sun-centered solar system entirely. In the Church's view, it would seem that politics played a more important role than science or theology.

The Catholic Church was able to tolerate Copernican theory in the abstract and even go so far as to calculate their calendars based upon it, but the church was unwilling to entertain the implications doctrinally. Galileo was excommunicated for his Copernican views and placed under house arrest. More than that, he was forced to publicly recant his findings. Still, Galileo never questioned his own belief in God. His daughter, a nun, received many letters from him that confirmed his faith, along with his love of the heavens he was discovering. Dava Sobel describes the journey of the devoted daughter and her brilliant and courageous father in the wonderful book, Galileo's Daughter.

What seems so tragic in the Catholic Church's early renunciation of Copernicus and Galileo is that, on a personal level, their studies of the heavens and the workings of the solar system only enhanced their appreciation of the majesty of God's world. These far-reaching scientists were not "doubting Thomases." They were men of godly and scientific vision.

According to the English calendar of his day, (41) Isaac Newton was born on Christmas Day 1643, in Lincolnshire, England, coincidentally within a year of Galileo death. It was as if the Italian scientist was passing the torch to the Englishman. Isaac's father had died just before the future physicist was born, and his mother left him with his grandparents on their farm. As a young man, Isaac was ill suited to farming, and others finally convinced his mother and family that Isaac would do well to pursue an academic life.

Isaac's senior year at Cambridge was interrupted by eighteen months of a school-invoked quarantine due to an outbreak of the plague. This might well have been the most successful school vacation in the history of western civilization. During that time, perhaps sitting under the legendary apple tree doing mental gymnastics, Isaac formulated his ideas about gravity, the three laws of motion, advanced the development of the calculus, developed an understanding of light in prisms, worked on optics, built a reflecting telescope of some forty magnifications and was otherwise a nuisance around the house. Many of those ideas did not come to fruition for decades. In time, of course, he would come to be regarded as one of the greatest complete scientists the world has ever known.

Yet his thoughts did not stop with science. He was a passionately religious man. It boggles the mind that he wrote more prolifically about religion and faith than he did about science. Science, he believed, was the handmaiden of religion. His life was a constant search for uncovering God's architecture. In his discoveries, he was entering the doorway of God's laboratory, but only the doorway. Newton remained quite humble about his observations. His was a faith that dared to look beyond the earth to incorporate the cosmos. God was the maker of a magnificent universe. Unlike some scientists, Newton held to the belief that God could intervene in the world, and did so, though Newton's own seemingly inexorable laws of nature would rule science for centuries to come. The universe, as Newton conceived of it, did not in any way dispel the belief in a divine and active God. Rather, for Newton, seeing simple laws that governed nature enhanced his appreciation for the Divine.

Maria Mitchell was the first American woman astronomer. She was born in 1818 in Nantucket, Massachusetts, to Quaker parents. Her father introduced her to astronomy, and she grew up with his words echoing in her ears: "An undevout astronomer is mad." (42) She became the professor of astronomy at Vassar College in 1865. Soon she became the Director of the Vassar College Observatory. In 1847, she made her name by discovering on her own what came to be known as "Miss Mitchell's Comet," (43) for which she was recognized internationally. The

Maria Mitchell Observatory on Nantucket honors her. In 1985, I viewed the rings of Saturn through the observatory's marvelous refractor. In her honor as well, a crater on the moon is named "Mitchell."

In addition to her passion for astronomy, Mitchell had a passion for ethics, was a suffragette in friendship with Elizabeth Cady Stanton, and had a kind of cutting-edge faith in the Almighty. While absorbing the Quaker-born belief in the intellectual equality between the sexes, she moved on to a more Unitarian understanding of religion and faith. She was starkly honest and believed that faith had nothing to fear from science. In her published lectures to her Vassar students, she wrote,

> "Let us have truth, even if the truth be the awful denial of God. We must face the light and not bury our heads in the earth. I am hopeful that scientific investigation, if pushed on and on, will reveal new ways in which God works, and bring us to deeper revelations of the wholly unknown."(44)

Not surprisingly, she found beauty and a reason to be in awe of the Almighty in her studies of the cosmos. In Beatrice Gormley's biography of Mitchell, *The Soul of an Astronomer*, she quoted the astronomer, "We especially need imagination in science. It is not all mathematics, nor all logic, but it is somewhat beauty and poetry. Every formula which expresses a law of nature is a hymn of praise to God."(45)

Albert Einstein, Nobel Prize winner for physics in 1921, was forced to leave his native Germany because he was Jewish. He emigrated from Germany to America in 1933 and became the professor of theoretical physics at Princeton. Einstein had a strong spiritual perspective, and because of his extraordinary status as a scientist, people, particularly in his later life, wanted to know his religious views. Did he or did he not believe in God? Einstein was usually clear but often misinterpreted. He usually expressed a great humility and sense of awe at the design of the universe, but he thought human beings could really have no definite knowledge about what God did or did not do. Because he had so much respect for the laws of nature, he did not think that divinity broke into them and acted here and there on behalf of humanity. He did not believe in a personal God, but he was in awe of the intelligent force that permeated the universe; he thought its design was beyond magnificent. In 1929 in an interview with Rabbi Herbert S. Goldstein, Einstein said, "I believe in Spinoza's God, who reveals Himself in the lawful harmony of the world, not in a God who concerns Himself with the fate and the doings of mankind."(46) And in an interview with *Time* magazine in 1930, he said, "I am not an atheist and I don't think I can call myself a pantheist. We are in the position of a little child entering a huge library filled with books, in many different languages. The child knows someone must have written those books. It does not know how. The child dimly suspects a mysterious order in the arrangement of the books but doesn't know what it is. That, it seems to me, is the attitude of even the most intelligent human being toward God." (47)

Einstein's view bears some similarity to the philosophy of nineteenth century naturalist, William Paley. Paley offered the famous analogy of the watchmaker: If there is a design, there must be a designer.

> "In crossing a heath, suppose I pitched my foot against a stone, and were asked how the stone came to be there; I might possibly answer, that for anything I knew to the contrary, it had lain there forever: nor would it, perhaps, be very easy to show the absurdity of this answer. But suppose I had found a watch upon the ground, and it should be inquired how the watch happened to be in that place; I should hardly think of the answer I had before given, that for anything I knew, the watch might have always been there. There must have existed, at some time, and at some

> place or other, an artificer or artificers, who formed [the watch] for the purpose which we find it actually to answer; who comprehended its construction, and designed its use. Every indication of contrivance, every manifestation of design, which existed in the watch, exists in the works of nature; with the difference, on the side of nature, of being greater or more, and that in a degree which exceeds all computation." (48)

Considering the different views of these icons of science, I am persuaded that, as theologians have long said, God is inscrutable. Yet, most of the great minds of history and science are in awe of the design of the universe, speechless at its immensity. Scientists find the idea that God may be personal even more challenging than most other human beings do. A personal God certainly cannot be known through the tools of science, only experienced through faith. But scientists generally, and certainly those I have mentioned in particular, are too humble to suggest that there is no intelligent design behind the universe's creation.

IN CELEBRATION OF THE FAITH OF SCIENCE

When I consider thy heavens, the work of thy fingers,
the moon and the stars that thou has ordained;
what is man that thou are mindful of him?

—Psalm 8:4

Man is not the center of the universe as once we thought in our simplicity, but something much more wonderful—the arrow pointing the way to the final unification of the world in terms of life. Man alone constitutes the last-born, the freshest, the most complicated, the most subtle of all the successive layers of life." (49)

—Teilhard de Chardin

What motivates men and women of science? What inspires them to become scientists? As with other professions, there may be many appropriate answers. I'd like to take a stab at a few possibilities. The men and women I've known who have become professional scientists are all imbued with an uncommon curiosity. It is more profound than the curiosity of a child who asks "Why?" in order to learn as well as to express himself ("Why is the sky blue?" or "Why do I have to go to bed now?").

The scientist's curiosity comes from a deeper place in the human mind and soul. It is linked with purpose and a yearning for meaning that centers him and grounds his understanding of his environment. It is more than just wanting to know how something works. It is more than mechanical tinkering. It often comes from wondering about the essence and perhaps the purpose of life. The work of a scientist, similar to the work of other professions, may often focus on mechanics and technology—how to make something or fix something or how to get something done. But the inspiration to do scientific work often begins with a quest, similar if not identical to the theologian's spiritual quest.

Fledgling scientists are indistinguishable from fledgling theologians in their passion to know the truth, regardless of where it leads. In the budding geologist, physicist, chemist, mathematician, astronomer, anthropologist, sociologist, psychologist, physician, artist, musician, or writer, there is often a burning desire to understand who we are as human beings. The passion of some leads them to strive to understand how the earth came to be and what may be its destiny. Some reach, in their curiosity and their work, beyond the earth itself.

Scientists shout "Eureka!" (not much different from the "Hallelujahs!" of many religious folk) at their discoveries. They appreciate the wonders of this universe, as did the biblical psalmists who are still lauded for their poetry and passion today. "The heavens declare the glory of God, and the firmament shows his handiwork." (50)

The passion of discerning minds, however, is often fed by skepticism. Out of skepticism, scientists have had the courage to question the status quo and reject some viewpoints that have been accepted as part of human history. Scientists seek to tell the story with greater accuracy than it was told by previous generations.

Many astronomers and astrophysicists begin exploring space in their youth when their idealism and curiosity is most keen. Their initial zeal is often driven by a profound wonder about the origin of the universe. As they deepen their investigations, they quickly discover that the universe does not yield its purpose and meaning easily. That is to say, space is impersonal. No matter how grand the power and light capacity of a telescope, the observer is greeted by a grand silence. While one may be in awe over the celestial beauty of a ringed nebula, or a cluster of thousands of stars, or a spiral galaxy that looks as if an artist painted it, a spiritual heaven cannot be observed. No angels fly past the observer's scope.

For many years, the deep silence of the universe was impenetrable, even with the most sophisticated instruments. The universe did not seem to communicate back to us. Recently astronomers have discovered different ways of listening to the universe. They interpret light's colors with spectroscopes; they penetrate formerly dark or hidden areas with x-ray telescopes; they listen to cosmic noises with radio telescopes. Some of those noises have apparently been present since the early days of the universe, perhaps even residual noises from the Big Bang itself. These noises, in fact, enable astronomers to map the universe more completely. In addition, scientists can glean the composition and age of the universe with much greater precision than before. Still, with all this sophistication, there is nothing revealed by those instruments that leads astronomers and astrophysicists to conceive of a personal God. They may be in awe of what they see; they may even discern a great design that hints at a cosmic Intelligence, but it is quite a leap to conceive of, let alone believe in, a personal God.

It is at this point that theologians, astrophysicists and cosmologists can have a meaningful conversation about the question of God's very existence. The theologian, also, has never "seen" the Creator. His training holds and his creeds testify that the God in whom he believes has been and always will be eternal and invisible. By faith, the theologian has also determined that this God, who is without material form, touches the minds and hearts of people as they relate to one another. This perception is not exclusive to human-to-human encounters. It includes all life and the inorganic world as well.

The theologian usually identifies this presence and power as the Spirit. Then he makes another great leap. In faith, he suggests that out of the darkness and the chaos, the "light of faith" illuminates the world, and connections to all life become possible.

Both scientists and theologians are attracted to light, the former as an energy that empowers the universe and the latter as a force that ignites the human spirit. The scientist appreciates more than most the attributes of light and its accompanying energy, which has the power to spawn life and to illuminate the Creation. When observing and contemplating the mysteries of the universe, scientists and theologians share a sense of awe and humility about our place in "the great vastness," and an enthusiasm for its magnificence.

The conversation between scientist and theologian has been going on for a long time. As scientists have explored the universe, new theories of its age and size have emerged. As recently as the nineteenth century, most people believed that the universe was but a few thousand years old. As mentioned earlier, men of curiosity and courage have since determined that the universe has been around for more than thirteen billion years. Today, most theologians happily accept this timeline.

That same sense of curiosity and wonder has also yielded information about the vastness of the universe. Not much more than a century ago, we thought the universe consisted of our galaxy alone, with our solar system at its center. So it is overwhelming to ponder that, by conservative estimates, there may indeed be more than one hundred billion galaxies. (51) Once again, scientists and theologians are linked by a sense of awe at this unfathomable vastness.

In the face of new scientific discoveries, we humans often fiercely cling to our understanding of nature. We fear losing our equilibrium and resist walking on new ground. In the face of this resistance, scientists must not only display courage; they must propose new theories that are nearly incontrovertible before most people, including people of faith, accept them. Scientists must have strong backbones when they put forward a new theory. They must first have the courage to test their theories and answer the challenges of their peers; second, they must then weather the suspicions of a hard-to-convince public.

How does new scientific knowledge affect people of faith? Some will protect the Scriptures and its literal reading. Some will assume an overprotective posture and push back against change, then only accept a new theory when the weight of scientific evidence is very high, and when they can see its value. One primary task of theologians is to articulate how new scientific knowledge does not adversely affect the ground of faith. Theologians can make the case that new knowledge often confirms and enhances faith.

Two examples follow: The first shows how the Church, in trying to protect "the Faith," reacted badly or slowly to new scientific knowledge; the second shows how theologians responded to new knowledge and enhanced the position of faith.

Consider the perils of the early-seventeenth-century Italian astronomer and physicist Galileo Galilei, who was put under house arrest during the Inquisition and forbidden in his day to discuss his first-hand evidence that, for example, the moons of Jupiter revolved around the planet. That single discovery was the most compelling visual evidence that every object in space did not revolve around the earth. Galileo went out on a limb and declared that Copernicus was correct. The Church reacted.

> "Unlike Copernicus, who postulated the theory as an hypothesis and so escaped condemnation, Galileo, in a later book in 1613, boldly asserted the Copernican theory and was at pains to point out its incompatibility with the interpretation of the Old Testament then in vogue. This earned him the condemnation by the Holy Office which issued a decree forbidding Copernicus' book 'until it was corrected' and prohibiting Galileo from holding, teaching, or defending the Copernican theory." (52)

By today's standards, the discovery that set all this in motion, the visual confirmation that the moons of Jupiter revolved around Jupiter itself was quite quaint. If a night was clear and Jupiter was in a favorable position, anyone with the privilege of looking through Galileo's small telescope could observe, in the course of a single night, one or two or perhaps as many as four readily visible moons revolve around the giant planet. The discovery was a nail in the coffin of the geocentric system, the theory that the Earth was the center of the Solar System and that all heavenly bodies revolved around the earth. It gave the new Copernican heliocentric theory some much-needed credence. But the information and new knowledge was, for a time, more than the Catholic Church could handle. Others had a difficult time assimilating this knowledge as well.

Common sense was the greatest enemy to a heliocentric system. Humanity, for thousands of years, had watched the daily cycle of the sun moving across the sky and the evening cycle with the stars moving from east to west. The commonly held views of a flat or round Earth, the center of it all, was crumbling, and people of faith and other scientists themselves were shaken by the notion. As a result, Jews and Christians feared that the biblical Book of Genesis would crumble with the new insights, and, perhaps, ultimately the Churches with it. It took the Catholic Church more than three centuries to forgive Galileo.

The second example is the discovery in 1947 of the Dead Sea Scrolls in the caves once occupied by the Qumron community in the century preceding Jesus. (53) These scrolls contained most of the Hebrew Scriptures. Prior to the discovery of the scrolls, people could naturally wonder if the Scriptures had undergone significant changes in the previous two thousand years. Could they have even been rewritten bearing little similarity to the earlier witnesses? The translation of the scrolls, among other things, revealed that the Hebrew Scriptures had not been altered. The Scriptures today read very much the same way they did two thousand years ago. In other words, one can trust their authenticity. This was a monumental triumph for the Bible. Skeptics thought that the various rewritings of the Bible had altered the text markedly and dismissed its current content as a result. But, the Hebrew Scriptures remain virtually intact from at least one hundred years before Christ to the present day.

The reaction to the scrolls themselves was varied. Jewish Scholars wanted to study them in isolation. Christian scholars had a hard time doing more than viewing them like tourists, and so the frustration of human nature attended the findings. But, germane to the purposes of this book, the Dead Sea Scrolls validated the dedication and skill of later translators. We can thoroughly trust that the Hebrew Scriptures we see today are as authentic as they were at the time of Christ. I personally had a chance to see the expansive scroll of Isaiah in 1967 in Jerusalem as housed in the Shrine of the Book in the Israeli Museum dome that contains the viewable scrolls. It was a magnificent experience.

In the long run, as with Galileo's trials, the truth eventually overcomes dark suspicions. Certainly, the courage of scientists and theologians is necessary for progress to occur.

Even the greatest giants of science have had to take risks when they exposed their religious or spiritual inclinations. Da Vinci wrote in code, and backwards at times to hide his theories from those who would, or could, cause him harm. In time, people embraced his views and universally saluted his genius. Arguably the greatest scientist of the seventeenth and eighteenth centuries, and one of the premier scientists of history, was Isaac Newton, whose thoughts turned increasingly to matters of faith as he matured. Yet, even he was thought to be getting too old and losing his scientific perspective when he turned his great mind to theology. It takes great courage for a scientist to express his faith.

The cosmos filled Albert Einstein with a deep sense of humility. He believed there was a spiritual dimension to the universe. When some suggested that the universe might just dissipate, expand into nothingness, he was fond of saying, "God does not play dice with the universe," (54) meaning there is divine design and purpose to the universe. While he was regarded with magical esteem, some people thought he should reserve his public views to science. Others were delighted to hear his views that touched on a cosmic spirituality.

With great respect for Einstein and many other scientists, I think it is important for people of faith to recognize the positive impact of scientists who are atheists. We Christians in particular often look askance at atheists, without recognizing the moral courage that many of them display. To reject the notion of God—while somewhat more socially acceptable today than in past centuries—is a stand for which scientists often have paid a dear price. In former eras, atheism was punishable in the harshest terms; in some fundamentalist cultures today, it is punishable by death. Many honorable scientists have found atheism the only position they can take with integrity. They may retain high moral and ethical conduct often associated with religious belief. Yet, in a posture I will call ethical atheism, the motivation to be good and seek goodness stems from an underlying belief, the product of diverse cultures, that the world is a better place to live in when we act in ways that benefit mankind, irrespective of religion or nationality. The honesty that many scientists display is rooted in their quest for truth and, in that sense, scientists are as morally motivated and grounded as many believers in religion.

Moral strength was evident in the lives and writings of such social scientists as Albert Camus, Friedrich Nietzsche and Jean-Paul Sartre. Because of the nature of their writing, they intentionally exposed their own atheism. As a "pure" scientist, I could add two-time Nobel Prize winner Madam Marie Curie to that list, who overcame a double prejudice at the turn of the twentieth century as a Polish immigrant and as a woman in a man's field. After the death of her catholic mother when Marie was young, she gave up religious practices. (55) Each of these strong individuals stood in a very lonely spot for a long time. Many great thinkers have needed and relied on faith though they may not have been attached to a religious institution. An ethical atheist must find that moral base within his own being and for his own reasons. He has much to teach the so-called "believing world" about courage itself.

How can we construct a bridge upon which people of science and faith can meet? Can this bridge include atheists, doubters, people of faith who do not subscribe to organized religion, and people of faith who are thoroughly involved in faith institutions? I believe the answer is "yes," but all the parties involved must want it to happen. They must bring to this bridge a willingness to listen and learn, and put aside that little bit of arrogance that infects all groups. In order to make this happen, people in either field of knowledge need to be well grounded in their own discipline and confident enough to entertain the ideas of other disciplines. The more secure one is in one's own discipline, the more one is free to open oneself to ideas that come from other fields of knowledge. Only the insecure become frozen in their way of thinking, exploring, examining, and experiencing the world. In fact, when one is secure enough, one seeks out what others have learned. Curious human beings thirst for knowledge. It is part of our evolutionary makeup. We want to expand, to understand more completely; we want to make deeper connections, to grow. An interesting example follows of a man who was secure enough in his faith and field of knowledge to be open to the thoughts of different cultures.

Some thirty years ago, I attended an ecumenical conference lead by a professor emeritus at the University of California and Harvard, Dr. Raimon Panikkar. The professor was a Hindu, Roman Catholic priest. Hinduism is a quintessentially accepting and inclusive faith. Dr. Panikkar was able to incorporate his Roman Catholic faith without losing his Hindu background. (56) I remember his instructive words, which now seem so far ahead of his time. He said something like this: "About twenty-five years ago we sent missionaries out into the world to save the heathen. About fifteen years ago we sent missionaries out and listened with interest to what different religious leaders had to say. About five years ago we invited teachers of different spiritualities to our country to teach us what they had learned about the Spirit."

Today, we need to foster the same kind of trust and collaboration among scientists and people of faith. We must value humility, and when this trait is mixed with a confidence that comes from having immersed one's mind and soul in a body of knowledge, collaboration is much easier. As is true of science and religion, and as sages have said, the more one knows, the more one realizes how much more there is to know.

Knowledge expands. Thank God for it. Without the "more," science and theology would come to a dead end. It is the quest for a richer, deeper experience that drives us on. Fifty-seven years ago, I first subscribed to Sky and Telescope magazine. At age fourteen, I could understand much of the content. Two generations and a lifetime of knowledge later, I cannot fully comprehend nearly as much of the total content as I once did. It is not merely that I am equipped to delve deeper than in my youth and therefore have more questions, but the field of knowledge needed in astronomy today has expanded enormously. To be on the cutting edge of the field today, one has to be conversant in half a dozen different disciplines.

One of the intriguing side effects of this general move toward complexity is the enigmatic truism about the acquisition of knowledge: The more that is uncovered or discovered, the more there is to be uncovered and discovered. We are not at risk of boredom in science or religious and spiritual disciplines. But even if boredom strikes, it is usually just a clue that we need to start digging again. Perhaps boredom is a signal from the Divine or the unconscious mind that we need to strike out in a different direction, to grasp what we are studying more deeply in order to reveal nature's wonderful secrets. But the secrets are revealed slowly. Humanity must work and be patient and to trust that the best is yet to come.

Men and women of science are motivated by the awe and majesty that they experience as the universe yields its secrets. And this sounds to me, a theologian, like a spiritual quest every bit as hallowed as my own.

THE BRIDGE OF LIGHT REVEALS THE GOD OF WONDER

The Heavens declare the glory of God,
and the firmament shows his handiwork.
One day tells its tale to another,
And one night imparts knowledge to another.

—Psalm 19:1

Science without religion is lame;
religion without science is blind. (57)

—Albert Einstein

One of the significant challenges to spanning a bridge between faith and science is finding some common ground in how each articulates the Divine or Original Source. Many religious people believe in a personal God. Scientists are often more comfortable speaking of energy and light, placing the idea of God outside their purview. While scientists and people of faith view one another with some understandable skepticism from opposite sides of an imaginary divide, it is ironic and edifying that the shared concept of light represents a bridge to closer understanding. With care, a bridge of light may be constructed from the building blocks of science and faith that closes the gap sufficiently so that a manageable leap across is possible. Let us see how this could be.

In the beginning, there was darkness, and then there was light. In the well-accepted "Big Bang" theory of our beginnings, the universe starts off with incredible energy. But it was without visible light. Gravity was so strong in the very early moments of the tiny universe that light could not escape. Only after the early universe expanded a bit, and the inward pull of gravity lessened, did light have the "strength" to burst forth from its shroud of darkness.

The religious expression of this phenomenon is not so different. Genesis begins in a shroud of darkness. Out of darkness, by the will of God, light emerged. "Then God said, 'let there be light;' and there was light…and God separated the light from the darkness." (58)

In Science, light is radiant energy. It is energy that can travel through space from one point to another without the need of a direct physical link. In other words, it can travel through space unimpeded. Light is described as either a wave, that is, a mix of electric and magnetic energy, or as a stream of particles, that is, photons or packets of energy. Their speed is phenomenal, reaching nearly three hundred thousand kilometers per second in clear space. (59)

Despite all of this light, much of the universe remains dark. It is estimated that dark matter permeates two thirds to three fourths of the known universe. Astronomers are attempting to peer into it, an ironic reversal of the creation stories. The darkness remains a huge mystery. The amount of dark matter may determine whether the universe is open or closed; that is, whether the galaxies will drift endlessly apart or, instead, reverse the expansion process when gravitational forces reassert an inward pull and the universe ends up in what is rather euphemistically called the big crunch. (60)

But that possibility is a long way off. In the meantime, the properties of light itself are still in the process of being understood. But much is known. Whether it is viewed as a wave or a particle, its speed in a vacuum is calculated precisely; it is refracted or bent slightly in different mediums; it can form a rainbow as it bends through the moist atmosphere following a storm; its color reveals the temperature and certain chemical properties of its source star. Discerning the spectrum of light in a star discloses its chemical makeup. The shift in the spectrum of light, known as the red shift, aids in calculating the distances of stars millions of light years from us. Basically, the red shift is a change in the color of light that occurs as space expands, stretching light waves toward the red end of the spectrum. The farther way a distant star is, the redder its light appears. The Red Shift is analogous to the better known Doppler effect when applying to sound, in which the pitch one hears rises when a train is moving toward you and lowers when the train moves off in the distance. With respect to light, when an object giving off light is moving toward us, the light is bluer; we say it is blue-shifted, and when the light is moving away, it is redder; we say that it is red-shifted.

Light opens up a great deal to us, but there is much more to light than meets even our twenty-first-century eye. There is little doubt that light will surrender more of its important secrets in the near future. At subatomic levels, new laws and new levels of reality about light await our discoveries. Today's scientists are suggesting that at subatomic levels, light's smallest particles or waves do not appear to pay attention to the laws of physics as we now understand them. While the speed of light has been well accepted as the maximum possible velocity for matter, some subatomic particles may not be governed by that boundary. As we learn more about light, we enter a sort of twilight zone where reality and mystery seem to merge. In that mysterious area, we enter the mutual realm of the cosmologist and the theologian. We could even say that in the mix there is a twinkle in God's eye.

In the Scriptures, light is prominent. The prologue of the Gospel of John heralds the gift of light. The gift comes directly from God. The Gospel uses the metaphor of light and life to express what is nearly inexpressible otherwise. "What has come into being in him was life, and the life was the light of all people. The light shines in the darkness, and the darkness did not overcome it." (61) Light and darkness exist together, but light (hope and life) are more powerful than the forces of darkness. In New Testament thought, the light that energizes us comes from God in the person of Jesus. In the Hebrew Scriptures, it comes with the dawn, (Psalm 130) "my soul waits for the Lord, more than those who watch for the morning." The new light each day energizes humanity and their spirit for doing His will. (psalm 148) "Praise the Lord from the heavens...praise him sun and moon..." Light as a gift from God and light as it energizes the soul are almost indistinguishable in the Hebrew Scriptures. (62)

What is being examined here is every bit as important as the big bang theory—it is the essence of life itself. It is one thing to describe how we came into being; it is another thing to address the meaning of why we are here. Science and faith are both needed to complete the story. The examination of physical light that enthralls scientists will help disclose much about the world of the future, and the light that gives life to the human spirit will be necessary to create life worthy of living throughout the planet. People of science and people of faith may rejoice in the parallels.

Addressing the meaning or purpose of our existence is at the heart of the field of theology. People of faith believe that God is the source of life and that meaning comes through their relationship with Him. Since God is not physical, that relationship flows through His Spirit and ours. Theologians, using metaphor and analogy, link light and spirit together. People of faith celebrate in liturgy and prayer the light that illuminates His handiwork. The psalmist declares that the universe displays through its beauty the glory of God. Similarly, the beachcomber stands in awe of the pink and purple hues of the sunset.

Science is more apt to describe light as energy. The source of that energy is the sun. Light and heat come from it, and without these entities, there is no life. The power that religious people call spirit the light from above, parallels energy in surprising ways. Spirit is not lethargic; spirit gives life and enables purpose. Energy gives life and enables purpose. Scientists are fond of studying energy in physical terms because their tools enable them to comprehend its physical properties. Those who pay attention to life and its purposes may find this spirit-energy in their relationship with others. Energy for healing, energy for moral decision-making, energy for loving the unlovable, energy for self-sacrifice, energy for working for justice—these are the spirit-making energies that abound on the planet among people of different faiths.

Both scientists and people of faith identify light as a key to creation and life; each explores light according to the tools of their discipline; one tries to understand its physical properties, the other, its spiritual ones. But there is enough connection between them to foster respect for one another, certainly enough compatibility to walk on the same bridge.

Since the Spirit is so critical to people of faith, and so easy to misunderstand, I will say more about it in the next chapter. All faiths speak of it but not with the same voice. I hope people of science as well as people of faith will embrace the following discussion.

THE SPIRIT OF GOD

Where can I go then from your spirit?
if I make the grave my bed, you are there also.
Where can I flee from your presence?
If I ascend to heaven, you are there;

—Psalm 139:7-8

As we could not call God a being, so we cannot call
him a person. He is not a heavenly, perfect person,
he is not a person, but he is not less than person,
he is the ground of everything personal. (63)

—Paul Tillich

The Spirit of God is the connective tissue that allows science and faith to engage in meaningful conversation together. Because I have used the term so frequently in this text, it is important to share my understanding of it, a Christian perspective. I hope that people from other faith perspectives will find that my thinking and faith resonates with theirs. I also hope that scientists and those who come from non-faith perspectives will respect this offering. When I refer to God's Spirit, I am referring to an invisible, immaterial power and presence that permeates all creation and is, in my belief, beyond creation as well. When I refer to the spirit within us, I am referring to the presence of that godly gift that connects humans to one another and to all other life. God's Spirit and the spirit within us are intimately and mysteriously connected especially when we are in the process of doing God's will; e.g. being merciful, granting forgiveness, working for justice, loving when it is most difficult, praying or meditating, listening to grand music and a host of other activities.

Theologians and other religious thinkers have long expressed that God is not completely knowable. Some go further and suggest that God is inscrutable. While this may be true on a philosophical level, there are some attributes and qualities of His that are strongly believed and felt.

In one of Jesus' sojourns into Samaria, he engaged a woman in dialogue at the local well, Jacob's well. In the course of their conversation, when she asked about God, Jesus said, "God is Spirit and those who worship him must worship him in spirit and in truth." (64)

First, notice that Jesus quickly affirms that God is not material; He is not a mega-being, human in form. God is Spirit. Second, Jesus links the Spirit and Truth. You can't have one without the other.

If we were to analyze that idea further, we would conclude that God is without gender or proportions or age. God is Spirit and eternal. We should have been able to get rid of that fleshly anthropomorphic God long ago. People with a scientific mindset can rest assured that theology, at its best, does not think of God in material terms.

Yet, in some respects, spiritual seekers don't want to give up the physical notion of God that humans have carried with them for millennia. We have a great need for a God we can touch. We need a God with whom we can relate. We want to feel the touch of the God of the cosmos.

That is, no doubt, why many of the great religions of the world celebrate a human hero. Some even deify him. Christians worship Jesus; Jews revere Moses; Muslims look to Mohammad; Buddhists follow the steps of Buddha; Confucians read the philosophy of Confucius; and so on.

For most mortals, the invisible, intangible spirit does not seem to be enough. We crave connection through the visual and the concrete. The Roman Catholic Church certainly understands this. Catholic churches are filled with statues of Mary and Joseph alongside Jesus, St. Francis in the garden outside and a crowd of saints for support. Jesus recognized that human beings needed a personal relationship with the Spirit of God. For many others, particularly those with a scientific perspective, images symbolizing spiritual ideas or concepts are not necessary. They are more likely to seek the spirit in silence or in the depths of theirs imaginations.

Paul Tillich was one of the great theologians of the twentieth century. In his book, *The Courage to Be*, he identified Divinity as "the ground of our being." (65) But as apt as that appellation may be, most of us have difficulty praying with any feeling to "the ground of our being." We find small solace in praying to a vague idea, an abstraction or to energy itself. We need the personal to engage our passion. Though Jesus told us that God is Spirit and those who worship him, worship him "in spirit and in truth," he also opened up another spiritual avenue for us. He provided us with a way of praying that made God personal. He invited us to call God, Abba or Father. So when we pray begin, "Our Father in Heaven, Hallowed be your name..." (66) In the word "Father," there is the invitation to the hungered-for personal relationship.

In that prayer, the Lord's Prayer, there is good news and bad. The good news is that this Spirit God can listen, comprehend and act. The bad news is that one cannot control the Spirit's actions. The Spirit is not a genie that can be beckoned to do one's will with the snap of one's fingers or even an elaborate incantation. Nevertheless, this Spirit God has more than enough capacity to be personal, and its power is awesome.

One of the powers of the Spirit is to be responsive to the voice of human prayer, not on our terms but on God's. Prayer has many forms and may be sought for many purposes. Those who are skeptical about prayer or communicating with a transcendent force may be attentive to this discussion as it requires no doctrine or denominational system. One may petition the Lord or the spirit of God but must be attentive and receptive to how the Spirit will answer. Prayer is a way to be in touch with God's Spirit—with the pulse of life at a level beyond the ordinary. Prayer places us in a mental/spiritual state that enables us to focus upon heaven's values, God's purposes. If one is quiet one can "hear" the still small voice, hear the music of the universe and feel connected to God's world.

One of the main purposes and benefits of prayer is to let go of our myopic vision, to surrender our self-centered point of view, to dampen the noise of our own rhetoric, and to awaken to the voice of the Spirit. It is then that "the still small voice" (67) can be heard and its message comprehended. Prayer is an attitude and disposition suited for few words and much listening.

Attentive silent listening is beautifully expressed in the words of early-twentieth-century Lebanese poet Kahlil Gibran in *The Prophet*:

> "And in much of your talking, thinking is half murdered. For thought is a bird of space, that in a cage of words may indeed unfold its wings but cannot fly. …And there are those who have the truth within them, but they tell it not in words. In the bosom of such as these the spirit dwells in rhythmic silence." (68)

A seemingly impersonal God has no trouble being personal with us. The Spirit is far more creative and communicative than we can imagine. A prayer posture ultimately requires a surrender of our desire to control the Almighty, the willingness to be vulnerable and open ourselves to a transcendent being. Prayer is not a game where we win by taking control, no matter how cleverly we disguise our words. Communication with the Spirit leaves deception behind.

But we are adept at deceit. Therefore, not many of our words reach the level of communication with the Spirit. We cannot cozy up to God and woo the Spirit into doing as we wish. So a discipline of silence and a disposition to listen is often a good avenue to successful prayer. Then the light often shines in prayer when we listen patiently. The spiritual light bulb goes off in a human temple swept clean of malignant thoughts and made ready for the presence of the light. The Spirit did not make us for darkness and death but for life and light.

The bridge of communication between the Spirit and humanity has been built spiritual brick by brick by millions of people throughout human history. It has been refined in the kilns of human struggles and pathos, human joys and sorrows, and the quality of communication has evolved over time, from very simple utterances to the Hallelujah chorus. Yet there remains, for every new generation, a chasm between the Spirit of God and human beings. No one has nor will probably ever be able to prove that God exists or is there for us. Human intellect always misses by an indefinite margin. A leap over whatever chasm remains is required. It is the leap of faith, a leap that requires riding on the wings of the Spirit.

Communication with the Spirit of God may be compared to a parent teaching a child to swim. At some point the child must plunge into deep water, over his head, and trust in all that he has learned in order to strike out on his own. Every part of nature seems to require this courage. The mother bird that nudges its reluctant baby out of the nest is a visible and common example. The baby has presumably watched its parents dozens of times, but when the time is right, the baby bird must fly on its own. It is an amazing moment of truth. There is no net to catch it, but with faith in its parents and courage all its own, the first flight is usually successful and the young bird soars into the light and into a new life.

This is the Spirit-induced leap that bridges the chasm between the human and the Divine, between darkness and light, and between non-faith and faith. Those who seek the affirmation of the Spirit in their lives must, at one time or another, make this leap. For many, this seems impossible, but the good news is that it has been accomplished successfully countless times. Some are theologians and other people of erudition, but many who make this leap are simple folk whose heart provides the energy to bridge the gulf. I might add that people of all faiths, all who seek a connection to the Spirit of God, make this leap.

In my seminary days, we called this the Kierkegaardian leap of faith, after the Danish philosopher Soren Kierkegaard, who so eloquently drew out for us the dialectic nature of faith and doubt. (69) In the end, after good science and erudite theology have enabled us to get to the chasm's edge, a leap of faith is still required. For those who decide to risk the leap, the fruits of life in conscious connection with the Spirit of God await.

THE UNIVERSE: AN INTEGRATION OF SCIENCE AND FAITH

How deep I find your thoughts, O God!
How great the sum of them!
If I were to count them, they would be
more in number than the sand;
To count them all, my life span would need to be like yours.

—Psalms 139:16-17

We inhabit a universe where atoms are made in the centers of stars; where each second a thousand suns are born; where life is sparked by sunlight and lightning in the airs and waters of youthful planets; where the raw material for biological evolution is sometimes made by the explosion of a star halfway across the Milky Way; where a thing as beautiful as a galaxy is formed a hundred billion times. (70)

—Carl Sagan, Cosmos

Carl Sagan, the splendid science populist, wrote in his book *Cosmos*, "The total number of stars in the universe is greater than all the grains of sand on all the beaches of the planet Earth."(71) He did not indicate whether he had done any measurements but suggested that there were about ten thousand grains of sand in a handful. I wondered whether he was making a hyperbolic statement about all the sand on all the beaches just to tickle our imagination. Was he merely trying to impress us with the mind-boggling number of stars in the universe? Most people, I think, would assume that in a cosmic competition between the grains of sand on the world's beaches and the number of stars in the universe, it would be a cakewalk for the beaches. His statement was laughable, except as literary license.

When I was teaching astronomy to high school seniors, I thought it would intrigue them as well as me to find out if Sagan's proposition was as preposterous as it appeared. So, I gathered an ounce of sand in a shot glass, and we counted the grains. In that batch we came up with approximately five thousand. We did a test sample with another ounce. The number was similar.

If you ever attempt to do this, you will notice several things: first, that grains of sand come in different sizes; second, grains of sand differ from beach to beach; and third, and this is the most egregious of all, grains of sand can be sliced into ever-tinier grains. Taken too literally, the idea of accounting for every grain of sand on the planet Earth falls apart. However, for purposes of this test, let's assume we are using an average-sized grain of sand, and say, five thousand grains for every ounce of beach sand. Then we come up with some amazing statistics.

To measure the sand on all the beaches requires some intelligent guessing. Intelligent guessing is also required to know the number of stars in the universe. But, it has been conservatively calculated by astronomers and cosmologists and affirmed by Sagan that there are no less than one hundred billion galaxies numbering an average of one hundred billion stars. Multiplying those numbers yields ten sextillion stars, or one with twenty-two zeros following. To give credence to Sagan's hypothesis, the number of grains of sand would have to equal that number.

Using the estimate of five thousand grains of sand to the ounce, we calculated that it would take six thousand tons of sand to yield one hundred billion grains of sand—about as many stars as there are in one average-sized galaxy. The one hundred billion grains would cover one small beach perhaps a mile long and several feet deep. But, there are one hundred billion galaxies to account for in this calculation. (More recent estimates double or more that number of galaxies.) There would have to be one hundred billion beaches a mile long to equal the number of stars in the universe. Even if the calculations are off considerably, Carl Sagan was more literally correct than most people would assume. It does boggle the mind.

I am not really concerned with the pinpoint accuracy of those figures, but more than any other example of the vastness of our universe, of which I am aware, this one makes the greatest impression and creates the most awe. When we consider the universe on this gigantic scale, we must at some point wonder who we are in relation to it. Are we important at all and, if we are, how do we know? What tools do we have to comprehend it? A theological perspective that embraces the universe enables us to address these questions.

On one level, the level of our material existence, we have made a habit of the illusion of our importance. From a purely rational, physical point of view, without the benefit of some spirit dimension in us, our material status within the universe is laughably insignificant. The illusion of importance seems inviting, for in strictly physical terms, it seems we have virtually no importance whatsoever. Even if we humans were to do the unthinkable and blow up our planet Earth, it would make not an iota of difference to the universe, no more than a single hiccup in the lifespan of a human being.

In a college physics course in 1960, I remember an impressive statement by Professor McKnight, a brilliant, MIT-educated scientist, who was not known for any particular spiritual perspective. During one lecture, when waxing eloquently about the universe, he stated that we should remember for the sake of our sanity that these wondrous stars of great size and length of "life" had no consciousness and had no idea about our existence. But, we humans, as miniscule and short-lived as we are, could consider the magnificence of the stars, the vastness of space, and appreciate the beauty of the universe in which we lived. Who would trade an hour of consciousness for millions of years without it?

The person of faith may be challenged by the notion of our smallness in the whole scheme of things. Faith must be willing to embrace the vast universe. We may not be able to comprehend it but we can be in awe of it. That can translate to being in awe of the Deity. It should remind people of faith that the God in whom they believe is no small time deity. The God of the universe is grand beyond our imagining. The vastness of the universe fills me personally with joy for being able to be conscious, if only a little bit, of this majestic Lord of Creation.

While the vastness of space can seem like a spiritual desert, one must remain open to the spirit of wonder if the music of the universe is not to be silenced. The universe yields beauty and meaning to the discerning and spiritually oriented mind. Though human life is very short relative to the "eternal" time of the universe, we can grow to live in harmony with it, identify with it and become one with it.

What about the future? Is it reasonable to dream of making contact and connecting to other life in the universe? We need to recognize that we are just beginning to take baby steps. While I approach this subject with enthusiasm, I note that just as I was writing this section, it was reported in the *Science Times* that renowned physicist Stephen Hawking suggested that we should be very cautious about space travel, as our history shows us that when we encountered new lands and other beings, we initially treated them with great hostility, as when Europeans first came to the Americas. Hawking's caution notwithstanding, the very local universe is, perhaps, within our reach in the foreseeable future, and given the chance we will go. The neighborhood of nearby stars is still light years away, but it is conceivable that connections can be made. If we seriously wonder whether we can make connections to other sentient beings in other galaxies, we must be prepared to encounter problems that even the most imaginative would have trouble bridging.

Seth Shostak, an astronomer at the SETI Institute, and author of *Confessions of an Alien Hunter: A Scientist's Search for Extraterrestrial Intelligence*, states, "Carting humans into deep space requires technology akin to wormhole rockets or matter-antimatter engines, the standard transports of science fiction." Shostak does posit a kind of virtual travel akin to what we do now with computers.

> "But there's another technology that's developing at a breakneck clip, and with which our grandchildren could make virtual trips to other solar systems. It's called telepresence—a collection of technologies that extends vision, hearing and touch far beyond the corporeal confines of our nervous system."(72)

Whether it is by chance or by higher design, whether the whole universe is accidental or overseen by God, the galaxies themselves are distant enough that human travel between them may never occur. If there is even the remotest chance of it in the distant future, it will happen long after the people of Earth have solved their most pressing problems and can engage global funds and resources cooperatively. I have a personal, unscientific theory that the galaxies have been designed to be at such a distance from each other that connections will not be made unless and until we humans on Earth have learned to live in peace and harmony with one other. We will either destroy ourselves before we possess that kind of galactic capability or we will learn to live in peace. If the latter happens, the future is long and open-ended. In the meantime, baby steps to the planets in our own neighborhood are necessary.

The isolation of the galaxies is not a trite matter. Consider what it would take to travel to our planet from another galaxy, to discover our planet amidst one hundred billion stars—a tiny, lightless planet some twenty-six thousand light years from the center of our own galaxy. Approaching us would be a neat trick also, as our sun is traveling about five hundred thousand miles per hour, its orbital speed in the outskirts of the Milky Way Galaxy. (73)

I should also mention the big chase that would be necessary in either getting to another galaxy or an alien spacecraft traveling to our own. For if one approaches from a distant galaxy, one has to first chase down the Milky Way itself, which is apparently traveling at about two million miles per hour through the cosmos. So a spaceship coming to us from another galaxy would be sophisticated indeed. In addition to the problems involved in an alien craft discovering and hitting such a fast-moving target, there is the problem of the time it takes to travel. The trip from the closest major galaxy, Andromeda, would take a million years or more even at the speed of light, a speed that in laymen's terms is greater than six hundred million miles per hour. (74)

So from anything we can currently imagine, travel between the galaxies seems like a remote dream. But human beings like to dream and perhaps, in the very distant future, we will overcome these travel restrictions.

If one enjoys playing God a bit, one could guess that in His infinite wisdom, He separated us—that is, the galaxies—by a sufficient distance, so that while we might destroy our own planet, even our own solar system, and down the road other planets in our own galaxy, we would be sore-pressed to hinder the lifestyles of other sentient beings in another galaxy. That perspective holds for beings in another galaxy as well. They cannot easily interfere in our lives. These various and multitudinous galaxies are for all practical purposes island universes, completely unto themselves. We are only able to view them telescopically and to imagine what goes on in them.

God, it appears, has "His cosmic hands" full, but He does seem to have a cosmic fussiness, a distant orderliness, and just maybe, He may allow unimaginable ideas to become possibilities as we become more godlike in our behavior and godlike in our celestial power. It would appear that while He has given us immense freedoms, there are limits for now. When I go out on a clear night, the cosmos seems to beckon me to dream of the myriad of galaxies and potential life in them. I smile, sigh and feel that this majestic God still holds the reins to it all.

EARTH'S PLACE IN THE GREAT SCHEME OF THINGS

Come now, let us reason together, saith the Lord.

—Isaiah 1:18 (KJV)

To me the universe seems quite the opposite of pointless. It seems that the more we learn, the more we see how it all fits together—how there is an underlying unity to the sea of matter and stars and galaxies that surround us. More and more the universe appears to be as it is because it must be that way; its evolution was written in its beginnings, in its cosmic DNA, if you will. (75)

—George Smoot and Keav Davidson

This book is an unreserved celebration of the material universe. It is equally a celebration of the spirit that, I believe, permeates the universe. The two, in my opinion, enjoy a kind of intimate symbiotic relationship. This chapter also celebrates that what we learn from science about the earth can be incorporated into faith. In it trust in God and science go hand in hand.

For the most part, we take our existence on Earth for granted. We rarely consider how delicate our existence is. We naturally see ourselves from the Earth looking outward, not from outer space looking inward, and therefore, we may feel substantial, significant and even important. We generally block from our conscious perspective how distant we are from our neighboring planets, how apparently lonely our planet is in the big sweep of things.

In addition, we are not usually aware of the amazing ways in which our planet is protected from lethal forces in deep space. In fact, Earth could not exist without an array of physical entities and powerful forces that keep us from being destroyed. For theologically minded people, it is Divinity that protects us. For scientifically minded people, recent discoveries illustrate the extraordinary celestial structures that provide for our safety and security. For most of human history we were barely aware of those outer bodies and forces that protect and enable our existence. Come with me on a journey in our solar neighborhood. Bring along your scientific interest and your faith.

The universe is very old. Our solar system is relatively new. During the first eight billion years of the expanding universe, there was no Earth. More than thirteen billion years after the mother of all cosmic explosions, (76) human beings appeared on this planet. What was God thinking, waiting so long for us!

The prevalent theory about our origins suggests that our sun came into existence some five billion years ago. It is a rather average sized star, but it is a rather perfect star when it comes to incubating life in its offspring, the planets. Our star has within it elements that it shouldn't have. Astronomers attest that while our sun is hot at the core, some 15 million degrees Kelvin, about 27 million degrees Fahrenheit, it is nowhere near hot enough to make the iron, gold, oxygen and many other heavy elements that are found within it. The core of the sun would have to be 100 million degrees Kelvin in its nuclear furnace to cook up those elements. (77) If our sun did not create those elements, what did? From where did they come? It is widely accepted in the astronomical community that the substance of our sun was baked elsewhere. Its original furnace was in an earlier star that was truly hot and very large. That early monstrous star came to a glorious end only millions of years after its own birth. It ended in a supernova, but it had done its job. When it exploded, it blew its contents out into the nearby universe. All the properties that now make up our sun were drifting about, waiting to coalesce into a new stable body. Our own sun did not create those elements; it couldn't have. They exist as the beneficiary of that earlier giant star.

It is necessary to our evolved life here that the sun is just an average-sized star. Stars such as our sun, which exist for billions of years, provide enough time for their planets to cool and mature. A few of those planets will be so marvelously placed in their solar systems that conditions for life as we know it may arise. Some stars are too big and some are too small to enjoy that possibility. A giant star burns out too soon. A small star doesn't provide enough energy. A nice yellow star like ours is just perfect. Our sun is medium-sized, but it has those necessary life producing elements within it, the gift of our parent supernova. When our gaseous sun formed, some pieces of it cooled significant distances from the main body, and went into orbit around the sun. Because of the way we formed, Earth is one of those planets that has elements including the heavy ones like gold and lead and oxygen that we need to be a viable life-giving planet. Some astronomers have dubbed it the *Goldilocks Planet*. (78) Just right!

The universe can be a dangerous place. It was especially dangerous during the early formation of our planets. Cosmic rays bombarded Earth, and they, of course, are deadly to most life including human life. Asteroids and wandering comets had the potential to knock our planet into oblivion. What would protect Earth and enable it to be the vital planet that it is?

Things have calmed down since our "birth" as a planet. As violent as storms can be now, they are nothing compared to those of the primitive Earth. The Earth sizzled for a billion years or so. Asteroids bombarded the planet as well. One was so large and the crash so monumental that a part of Earth was sliced off, and the remnants formed and cooled and became our moon.

Our moon is not just a lovely object for stargazers and lovers. It helps to balance our planet. Without it, Earth would wobble out of control. Later, it rained on Earth for thousands of years and the cooling process began. Most importantly, oceans began to fill up much of the planet and remained in watery form when the planet had cooled enough. With all those precious elements, cooked earlier in the supernova that gave birth to our sun and planets, now safely within the core of the earth, the heavier ones sank to the interior and provided a moderate gravity for our planet. Earth was ready for life to emerge. That happened a couple of billion years ago.

In 1994 a major comet came toward Earth from the distant regions of the solar system. It was observed well before it crashed and provided the world with quite a show. It is now called Comet Shoemaker-Levy 9, (79) after the scientists who first spotted it. It did not crash into Earth. It crashed into the largest of our planets, Jupiter, a planet eleven times the diameter of Earth and more than three hundred times its mass. Like a machine gun, Shoemaker-Levy 9 hit Jupiter, rat-a-tat-tat, at least nine times, each time resulting in a monstrous, cavernous hole in Jupiter's gaseous surface. One hit created a hole in the planet as large as our entire Earth. As if Jupiter were a cosmic whale, it gulped and "digested" the comet fragments. Within months, few discernible traces of Levy-Shoemaker's horrific bombardment remained. Jupiter saved us.

How many times had Jupiter saved us before? Considering that from our tiny earthly window seat in 1994, astronomers observed this with mouths agape, we can surmise that Jupiter has saved us thousands of times before. Jupiter is not the only planetary savior out there. There are other giant planets, Saturn and Uranus and Neptune, each farther out than the other. Each of them has taken cosmic hits. Uranus, four times the diameter of Earth, was once knocked over on its side by more than 90 degrees. What a colossal impact it must have been to knock that giant planet so far off its original axis!

Other forces help to protect the earth from interstellar radiation such as harmful cosmic rays. "The solar wind, a tenuous flow of mainly hydrogen and helium that sweeps across the solar system," (80) is one such possibility. Also, surrounding the solar system is a cloud of rocks, gasses and unknown bodies. Beyond Pluto, millions of tiny bodies exist called the Kuiper Belt and beyond that lie trillions of particles and small bodies called the Oort Cloud, named for twentieth-century Dutch astronomer Jan Oort, who first proposed its existence. (81) This area of space is teeming with trillions of dust particles to small asteroids. It may be that the Kuiper Belt and Oort Cloud help to protect us from cosmic rays, though that is yet to be proved.

Some four billion years after the formation of Earth, human beings came into existence. We were preceded by ages and ages of life. Like a good Thanksgiving meal, Earth took a long time to prepare itself for the potential feast it now represents. Whether it is oxygen and water, iron and gold, oil and diamonds, or an agreeable temperature for life—all these vital components were necessary to enable life, as we know it, to have a chance.

A theologically minded person believes the hand of God created the universe. He or she bows in awe to this God of wonder and power, this God who so majestically orchestrated our possibilities. By contrast, many scientists see the signs of evolution in the universe itself. Astronomers sometimes call us "star children," as every molecule that makes up our bodies originated within stars. This is a wonderful, light expression, and it tells us much about our beginnings. But, for me, it does not compare to the powerful connection to the universe that people of faith offer in the same context—to be a "child of God." How much grander it is to believe that we are children of a loving Father. A "star child" is not loved, though he was marvelously molded. Is it not more wonderful to be loved by the Author of the universe and to be given a purpose and a destiny? It so happens that the two expressions, "star child" and "child of God," each contain some truth. Neither perspective should be ignored. The term "star child" lets us know about how "our ingredients" came to be, and the term "child of God" gives a hint about our purpose and destiny. Each is needed to help us know who we are.

Science and faith may link up in a great new story or rekindle an old one. The story today cannot be told without respect for the discoveries of science, nor can it be told in all its glory without considering faith in relation to Earth's most amazing creation, you and me. This is truly a love story, between the Creator and the Creation. Viva la science that gave us this new story, and viva la faith that helps us to integrate it and give it meaning!

MIRACLES OF FAITH AND SCIENCE

Soon as the evening shades prevail,
the moon takes up the wondrous tale,
and nightly to the listening earth
repeats the story of her birth:
whilst all the stars that round her burn
and all the planets in their turn,
confirm the tidings as they roll
and spread the truth from pole to pole.(82)

—Paraphrase of Psalm 19:1-6

The most incredible thing about
miracles is that they happen. (83)

—Father Brown Omnibus

This short chapter introduces the subject of mysteries that we encounter in the far reaches of space as well as on Earth. Part of our human experience is the appreciation of mysteries that overwhelm our ordinary sense of cause and effect. Some anecdotal experiences that test the limits of coincidence will follow in the next chapter.

We begin with sights that the Hubble Space Telescope has revealed. It peers out into space from its lofty height, nearly airless atmosphere, a couple of hundred miles above Earth, circling, searching and seeing with amazing clarity distant objects never before seen by human beings. As it opens the universe for our perusal, we stop and ponder the pictures it sends to Earth; our minds reach for understanding and meaning as new matter comes into view. Much is mysterious, even to the skilled observer.

Without the impediment of the earth's atmosphere, light from distant space fills the telescope's mirror. There are patches of impenetrable blackness, dark matter, which still mystifies us. But, with the aid of light, we have much to ponder and discern about the universe—its strong and weak forces, whether or not gravity behaves the same throughout, whether space is curved or flat, the relation between time and timelessness. The pictures from the Hubble Telescope invite a myriad of questions from scientists and laymen who view them. They create a sense of awe in everyone.

In the sweep of the telescope, nearly by accident, we encounter beautiful structures in deep space. Some pictures suggest to us comparisons to earthly images such as a ringed nebula or diamonds in the sky, or the "eye of God," or images of the birthing of stars. Yet we rarely call these awe-inspiring deep-space wonders miracles. When we do, it is usually in a manner of speaking, as when we exclaim, "Oh! What a miracle!" That sense of awe happens when we glimpse a sunset from an expansive beach, or from the peak of a high mountain, or from a wide-open prairie.

When we feel overwhelmed by exquisite beauty, we feel free to surrender and exult in the word "miracle." Pictures from deep space can yield that feeling; perhaps we are secretly proud that we live in such an awesome, amazing universe, beyond our most profound imagination.

Most laymen ask from time to time whether everything is governed by the immutable laws of nature, that is, does everything proceed according to fixed and forever rules of the universe? That leads us to wonder whether divine power could intervene in those laws. Are the only miracles the "miracles of nature," that is, extraordinary happenings without the benefit of recognizable consciousness? Are miracles merely the magic of coincidence?

The scientist cannot confirm nor corroborate anecdotal testimony to "miracles" because they are not subject to experimentation or replication. A second person may not see the same thing the next day. One cannot make a hypothesis of a miracle and experience it anew. We creatures are certainly happy with that. We want to and do trust that gravity will be constant, that when we jump from a two-foot height, we will always fall at the same rate of speed. The laws of nature prevail. Nevertheless, the question lingers: Are there times when the laws of nature are upended, suspended, overturned or ignored? Does God act outside these immutable laws, deus ex machina? I am wary to state so, and yet I believe miracles happen all the time. The laws of nature are not yet completely understood, and God may smile about our skepticism.

I am very attracted to something I call divine coincidence. It is not the overturning of the laws of nature. It is more like a spiritual force that acts with or reacts to the laws of nature. Consider a situation that is altered in such an amazing way that injury is avoided and a person is saved from an otherwise harmful or deadly consequence. Consider the situation where a person tries desperately to make his airplane flight, but some unusual happenings repeatedly thwart his efforts, and the flight he just missed crashes. Some would call this ordinary coincidence; those to whom it has happened would be likely to call it God's grace. I have experienced in my own life and heard many testimonies about extraordinary occurrences where divine coincidence (mind-boggling luck) had a saving grace about it. Others who hear miracle tales may assign them to luck—wonderful, improbable luck. People with a more Eastern interpretation might say fate determined the outcome, or one's time wasn't up yet, or a myriad of interpretations of out-of-the ordinary phenomena. Yet, an air of mystery lingers. When such an event happens to you, the conviction of that mysterious presence is difficult to shake.

Some might ask: Do we have a power within us, an intuition, or another sense that warns us or even directs us to take a path different from the one we expected to take? May we communicate this sense to others? Do we have a karma that, under the right circumstances, informs us of foreboding possibilities and in so doing we are rescued? Are there attributes of the brain, mostly untapped, that on occasion grant us a greater power than we ordinarily use? Do some people possess a sense of the divine, a spirit that yields knowledge beyond the rational, verifiable and visible? Are miracles made within us or within others? Some of these questions are beyond the scope of this book, but they point out possibilities where the mysteries of science and faith touch shoulders and perhaps overlap. Some of these questions move from the philosophical realm into real life experience in the anecdotes that follow in the next chapter. As you read them, you might hold the following questions, as well, in your minds.

Must science and faith completely disregard or demean each other because of the propensity of the first to deny the possibility of miracles and the propensity of the second to insist upon it? Could there be agreement to disagree and not disparage the field of one or the other? Can we rejoice in letting science be science and faith be faith? Can we be thankful for the gift each discipline offers?

In my life, there have been many occasions when my conscious mind did not contribute an iota of knowledge, but where an action of mine, perhaps driven by my unconscious mind or some positive spiritual force or intuition, saved a situation. Over many years, I have learned to trust that whatever unknown force is at work, I am very grateful for its presence. I have no problem calling this unknown force the spirit of God. But name it or not, in my experience, the force is real.

In the following chapter, I will share six stories that have been part of my experience. Either they happened directly to me, or they were told to me firsthand from people I know well and trust thoroughly. Because these stories are my experiences, not yours, it is natural for one to be skeptical. Yet, I risk telling them in the hope that for a moment you will suspend your skepticism and let them simply live.

Before relating those stories, I want to tell a personal one where I felt the beckoning of the spirit but I did not heed it to my regret. Again it was 1994 and I was acting as interim minister at St. James the Fisherman in Islamorada, Florida. On a Friday in April, I drove my wife to Fort Lauderdale to catch a plane to visit her mother. I was to conduct services on Sunday so I drove back to the Keys. But as I got on the highway I had a strong urge to see my father who was dying in Sarasota, across the state from Fort Lauderdale. So strong was the urge that I pulled off to the side of the road and wrestled with these impulses. It had been suggested that my dad had about a month to live. I had planned to drive up and see him on the following Monday. So I pushed away these very strong impulses. I rationalized that I needed to get back to the Keys and reasoned that there would be time enough to see him next week. I received a call from my mother on Saturday that my dad died. I had missed the opportunity.

MIRACLES CLOSE TO HOME

O sing to the Lord a new song, for he
has done marvelous things.

—Psalm 98:1

The modern mind always mixes up two different ideas: mystery in the sense of what is marvelous, and mystery in the sense of what is complicated. That is half its difficulty about miracles. A miracle is startling; but it is simple. It is simple because it is a miracle. It is power coming directly from God instead of indirectly through nature or human wills. (84)

—Father Brown Omnibus

The first story is deeply rooted in my family history. I was not an eyewitness, but I certainly was a beneficiary of the experience. To capture the essence of this adventure, I have to share a peculiar characteristic of my mother. As a born-and-bred woman of Manhattan, she never left the apartment except with a purpose, that is, for groceries or shopping or other such purposeful occasions. She never, ever just went for a walk. Even when we later moved to suburbia, I can attest that she never went for a walk, and if she were to visit a neighbor two hundred yards from our home, she drove.

The Apartment

It was the winter of 1938, before my sister and I were born. My mother and dad were married, living in Manhattan. The apartment was on the fourth floor. It was mid-morning and my father was at work. He was in investments and worked for the Springfield Fire and Marine Insurance Company. On this morning, my mother was home alone, ironing some clothes. With no conscious purpose or destination in mind, she went to the closet, picked up her winter coat and left the apartment on foot. She walked for about forty-five minutes.

When she returned, the apartment building was filled with smoke and part of it had collapsed. The fire had already made the news, and my grandmother heard the radio report. She called my father at work, got him out of a meeting and told him what had happened. My mother found my dad, along with firemen, going through the rubble, looking for her. Shaking with relief, my father was everlastingly grateful.

When the full story came to light, it turned out that a man in the apartment two stories above my mother and dad's had turned on the oven, stuck his head in it, and committed suicide. As the gas expanded, it caused a mighty explosion that partially destroyed the apartment building. My folks' apartment was ruined, but my mom was safe.

I have, of course, had many years to think about that story. It is entirely possible that we possess some sort of intuitive sense, possibly God given, such as that which protected my mom, something not of her conscious mind yet powerful enough for her to eschew her normal habits and respond uncharacteristically. It could as well have been a "divine push," or the two could be connected, but it was not just a stroke of luck. That would stretch my credulity too much.

The second tale of deliverance was not quite as dramatic but is as meaningful to me as the first. It, too, involved a similar kind of divine push or extrasensory power.

The Metal Swing

In the spring of 1973, I was a relatively new parish rector and, more importantly, a proud and happy young father of two little girls, aged just five and not quite four. I came home one afternoon to find a small impromptu gathering of kids.

We had a big metal swing set in the back yard, complete with a seesaw, some climbing apparatus and the pièce de resistance, swings with seats facing each other. The swings were connected by a metal floor. Three small children could sit on each seat. However, there were seven children at this gathering. My daughter Chris, the smallest, stood on one of the seats between two kids, seemingly small enough and tucked in enough to be safe. The crew of happy kids was unable to move the swing by their own collective power, so Dad, of course, became the engine that could.

To power the double swing, I stood on the crossbar that braced the swing set closest to the seats. My feet were about four feet off the ground, and I began to move the lively load of little people. Everyone was laughing, the swing got higher and higher and, with one extra movement, little Chris, in a nanosecond, flipped backward over the swing and landed on the back of her head. As she lay on the ground, the metal swing set moved back and forth over her head, missing her face by an inch with each pass.

Stopping the swing as quickly as possible, I lifted her carefully and brought her into the house. She was unconscious. I shouldn't have moved her, but I did. I cradled her for a time on the sofa, and, of course, prayed as never before. She awakened after a few minutes and seemed okay. All it amounted to was a bump and bruise and a brain rattling.

This story would be like so many other stories of the trials and tribulations of parents and children were it not for something that had preceded this scary adventure. The day before this event, I was out in the back yard and, for the first time, I looked carefully at the swing set, purchased a year earlier. I noticed that the floor of the double swing seat, where several could rest their feet, was just a couple inches off the ground. I decided to hoist the four large metal posts of the swing set and place them in a slightly different location. The double swing seat's metal floor now stood about ten inches off the ground. Why I chose that moment to do this, I am not quite sure, but I am eternally grateful. I certainly was not consciously envisioning any danger. I think I did it because it appeared to me that if the posts sank any deeper, the double swing seat floor would not clear the ground, but I was not in the habit of being so observant. On that day I was.

After I had time to reflect upon the incident, I realized that if I had not moved the posts up the day before, the metal floor of the double swing seat floor, weighted by six children, would probably have mashed her face as it swung back after her fall. As it was, it missed her face by an inch or two.

This is the kind of tale that many of us have known in different ways. It may appear to some as just a spectacular piece of good luck and, at the very minimum, it was that. When this kind of event happens to you, however, you are humbled. Whatever your disposition is, you are grateful, and, for a time at least, you thank the hidden powers of the universe and, perhaps, your God for the wonderful outcome. I know I did.

The laws of nature were not for a moment interrupted. They functioned just fine. What was surprising and marvelous was the serendipity or divine timing of moving the swing set. This, of course, invites more questions if one considers that the outcome was not just pure luck. For instance, if some consciousness moved me to hoist the swing the day before the event, then in some way was the actual incident foreseen? That question may invite inquiry beyond the scope of this book. But the happy outcome to this small family saga suggests that another dimension of reality, perhaps one that we take for granted, intervened on that day.

The next "miracle" story is one that I might call the victory of unadulterated enthusiasm and positive faith over cold technology and human angst. No laws of nature were broken in this drama either, but the laws of nature certainly clashed with a family that forgot for a split second to pay attention to them. On Earth and out in space, the laws of nature are often cold and impersonal, without compassion. You rail against them at your peril. Yet, the human spirit can soar above them, and sometimes in the soaring process, human beings can laugh at those inexorable laws.

Madeleine and Chloe

It was the winter of 1996, a cold and clear Saturday afternoon in January. The setting was a small parking circle near the chapel on the Salisbury School hilltop, a private school in the Connecticut Berkshires where I was the school chaplain. The players in this saga were faculty parents Marcia and John King, their dog, Chloe; their son, Josh; the headmaster's wife, Sally Flood; and her dog, Madeleine. I was basically an observer whose role became a chronicler who wrote of and told this story on several occasions.

The Kings were running late. They were going to a wedding a couple of hours away, but one thing or another delayed them. Now there was no time to waste. They lived in town, but needed to drop Josh off at the school. John usually drove his own car, a Toyota, but on this trip he was at the wheel of his wife's Dodge Caravan. While parked in the circle, John realized he needed to get something out of the trunk to give to Josh. As he was opening the driver's-side door, he hit a lever (thinking that he was in his Toyota) that he thought would open the trunk. Josh had disembarked and Marcia, also, was outside ready to hug her son goodbye. With that click of the lever, all the doors locked. With John's keys in the ignition, the car was still running, and Marcia's purse, with her keys inside, was resting oh so comfortably on the front seat.

Immediately, the nastiness of the situation became apparent. John lost it for a moment, kicking the tires. Each parent looked at the other with accusatory eyes. They would be late for sure. The maintenance men at the school had gone home, and "Murphy's" sardonic laughter could be heard from the depths. Leaving my chapel office, I happened upon the scene at that very moment. I suppose my presence prevented a family explosion, but the real angel of the day was only a hundred feet away.

Irrepressible Sally Flood was walking her middle-aged basset hound, Madeleine. As she approached, she could sense the consternation and the Kings' plight. Sally, ruled by an enthusiastic spirit, lost little time in lamenting. She immediately picked up Madeleine. Inside the Caravan, on the front seat, also sat the Kings' very-much-alive springer spaniel, Chloe. Sally quickly pressed Madeleine's nose to the window. Chloe immediately went berserk, jumping up at the window in frantic delight or in sharp warning, possibly a combination of canine emotions. Within a minute, Chloe, jumping all around the right-front window, hit the automatic window lever, and, voilà, we all heard the glorious sound of the window rolling down. The day was saved.

It was wonderful timing after all. Just the right moment for Sally and Madeleine to make an appearance, just the right enthusiasm, and just the right bounce in Chloe to affect this happy ending. John and Marcia's angel had appeared, and they made it to the wedding on time. The Kings would laugh and celebrate this event for years to come.

I cite this example for a few reasons. Some, perhaps those of a scientific bent, might have thought of other ways to unlock the car. Those with creative rational thought would have considered rigging a wire hanger, or calling the police, or calling a garage located several miles away. However, those potential solutions would have taken too much time for the harried wedding-bound couple. Because of the urgency, some might have considered breaking the window. But enthusiasm, irrepressible will and vibrant faith in a positive outcome, as well as the "luck" of exquisite timing, resulted in solving the problem faster than most "logical" methods. And certainly the saga provided a story better than any other. Something about Sally's energy and attitude enabled Madeleine to excite Chloe just enough. Sometimes, the capacity to remain positive in the face of discouraging situations enables "little miracles" to happen. When we don't limit ourselves, we are able to see what others may not.

The fourth short story moves us beyond the human-to-human experience of serendipity into an arena that includes some of the animal kingdom, certainly at a deeper, more intentional level than the previous story. It is not so much about divine timing as it is about harmonious feelings that can be generated between the species in rather miraculous ways. This story comes from Barbie Riegel, a friend and respected member of the vestry in my summer parish on Fishers Island, New York. I will let you hear it in Barbie's own words, from an email she sent in September 1999 about an incident earlier that year.

Weesie's Dophins

Dear Steve,

As promised, I must tell you of the wonderful Bermuda dolphin/porpoise story.

Weezie Outerbridge, aged 35, was the daughter of great friends of, not only Jerry's [Barbie's husband] and mine, but was also extremely close to Dick and Sally Flood [the same Sally as in the previous story]. They all spend their summers together in Jamestown, Rhode Island.

Weezie had returned to Bermuda four years ago to teach at the private Bermuda High School for Girls. She had a phenomenal way with kids, and, having grown up in Bermuda, she loved many sea creatures. But in particular, she loved dolphins. She had made them her focus for years, her symbol, her mascot, so to speak, even though dolphins have not been seen in Bermuda waters for many, many years. On the day she died in Bermuda this past April (she was only sick for a month and a half; cancer was galloping through her system), it was reported that many Bermuda fishermen, who fished off the south shore that day, came back just overwhelmed by the sight of a huge pod of dolphin [sic] (well over 60 of them was the estimate) frolicking and playing in the waters just off the south shore.

And this summer, when her younger sister, Liz Outerbridge, and her beau, Larry Richardson, became engaged, they were out in a boat off of Virginia Beach, where his family lives. That night, a single dolphin swam around the boat, circling it three times, then went up on its tail and gave her a big wink and swam off! (85)

Was this a mystical send-off from Weezie? Each of you may interpret this differently but the specter of mystery from the deep and from heaven is there to ponder.

I include this letter in the conviction that many readers will recall and be able to reaffirm events in their own experience that somehow parallel this story. Perhaps tales like this will help us give voice to the mysterious power that breaks human barriers and is even present between species.

I move on to an incident that appears to exceed the limits of ordinary human spirit or intuition. It has divine help written all over it. It remains in that category we call mystery, but it cannot be ignored. Certainly it permanently affected the recipient of this "miracle."

Jeanie's Pool Angel

When my wife, Jeanie, was a little girl, she experienced a life-saving event. I asked her to write it down in her own words:

> It was the summer of 1950 and our family had moved that spring during my first grade year from Caledonia to Galion, Ohio. I was seven years old. My sisters were eight and twelve. Each day during the summer our mother dropped us off at the town swimming pool. This is what happened to me on the very first day. The pool was already crowded when we arrived. We waited in a long line, paid our money, and were given a basket for our clothes. Because of the polio scare, I had not yet learned to swim. Being in a crowd and also swimming with others were thought to be ways in which one could pick up the virus.
>
> The older kids organized a game of follow-the-leader. It involved jumping into the pool quickly at deeper and deeper depths. Soon we were beyond the shallow end and its warning ropes. As the littlest kid, I jumped into the deep end last. The others were on to the next jump. I sank like a stone, and almost immediately was on the bottom. Suddenly a little girl, about my age, appeared at the bot-

> tom of the pool with me. With her arm between my legs she pushed, with uncommon force, off the bottom, and I found myself at the side of the pool looking at her. I sputtered, "You saved me." She never answered, looked at me with what I remember as a questioning look and disappeared. I didn't see her again. All that summer I looked for her. I never saw her again. My sisters saw none of this, and had a hard time believing me. I didn't tell anyone else of this until 35 years later. Who was this little girl who knew so much about lifesaving; this angel out of nowhere never to be seen again?

I have heard many angel stories over the course of my life—those messengers of good will who appear at critical times. I do not doubt the stories. I cannot explain them, but I do look heavenward when I am told of them and hear in my mind the chorus of angels.

The sixth and final story in this chapter also involves timing, but also appears to involve a force that is counter to the laws of nature. I do not know how to explain it. I do know that this happening awed me, and that I am not ashamed to tell the tale. I was there and, along with others, witnessed the following situation.

The Backward-Ticking Clock

It was the spring of 1975. The Vietnam War had just concluded and some South Vietnamese immigrants had made their way to this country, supported by local town groups. Such was the case in Collinsville, Connecticut, where I was the rector of Trinity Episcopal Church. A Vietnamese man whom we were supporting had been an air traffic controller at the airport in Saigon and was among the last to escape as the North Vietnamese were taking over the country. Although he was a highly intelligent, highly trained man, he was willing to start from scratch in America. My role was to take him on some job interviews. The first one was as a handyman in a nursing home in Simsbury, Connecticut. Our appointment was at 3 o'clock on a Friday afternoon.

It turned out to be payday, and the nurses were all reporting to the manager's office to receive their paychecks. We were told to report to the same area. We had arrived early and sat in the room in which all that flurry of happy activity was happening, ten feet in front of the comptroller's desk. It was about 2:40 p.m. There was a large clock on the wall that everyone could see. The nurses continued to file in to be paid, and the clock ticked away. When it was about 2:55 p.m., the clock momentarily stopped and then inexplicably began to tick backwards. Soon it was ten minutes before three and a little later it was 2:45 p.m. Without fanfare or comment the Vietnamese job recruit jumped up, grabbed a chair, moved it under the clock, stood on it and removed the clock. He fiddled with it for a couple of moments, put it back, and the clock ticked forward as a proper clock is supposed to do. The nursing home comptroller saw that bit of initiative and ingenuity and hired him on the spot. He accepted, and that was the end of the interview. I went home with my jaw on my lap—deus ex machina if I ever heard of it! It seemed to me that, for a moment, the laws of ticking clocks, if not the laws of nature, were broken or suspended. I have no rational explanation for this event.

There are times when the appreciation of mystery is what we are granted. There is no retracing the evidence to support our experience, no instant replay. Merely to tell the tale is to invite doubt. That is one reason miracle stories are not told more often. They invite skepticism and worse. Nevertheless, this is my personal testimony. Consider the forces that had to come together in this experience. A little initiative on our part, a clock, special timing in several senses of that word, an alert job recruit, an attentive comptroller, a bunch of witnesses, and the mystery of divine timing that can't be easily explained.

Scientist or theologian, we see and appreciate amazing wonders. We interpret them differently. Scientists may witness any of these occurrences, and their curiosity is aroused. They will immediately jump to find a logical cause. The Vietnamese man was not distracted by the mystery of the occasion, only intent on a way to solve the problem; he had the logical can-do mentality of a good mechanic or scientist. But all humans can also appreciate mystery even as they go about solving problems. It is the divine timing of events that causes most of us to exclaim, "Eureka" or "Hallelujah." The response of the scientist and the response of the one who appreciates the mystery are not antithetical. They are complementary. We humans have capacity for each. There is nothing that prevents us from being both scientist and theologian.

I have presented six stories from my experience or that of others close to me. While the stories are related, they are slightly different as well.

The first two experiences (the exploding apartment and the metal swing) suggest an exquisite timing and saving power that interrupts a potential danger. This source of protective power may come from an unidentified energy within us, or what we often call intuition or extrasensory perception; or the power may come from without, from a divine source. Perhaps these powers commingle from time to time.

The third story (the two dogs, Madeleine and Chloe) suggests that extraordinary things can happen when the spirit of enthusiasm meets with irrepressible faith.

The fourth story (Weezie's dolphins) suggests a species-to-species connection that transcends space and maybe time. And the connecting tissue may be in the power of love among the individuals.

The fifth story (Jeanie's pool angel) suggests that when confronted by imminent danger to life, other forces within us and perhaps outside us have the power to transform a critical situation. A sense of mystery surrounds this one.

The sixth story (the backward-ticking clock) takes us decidedly outside the intuitive and the enthusiastic. There is mystery here, but to someone who is open to a divine nudge, it seems more logical to accept that something outside of human control was operating.

I am reminded in the telling of these very personal accounts of an ingredient that seemed to be present in most of the miracles attributed to Jesus in the New Testament. Most often, though not always, at least in the texts as we have them, his miracles involved the faith and commitment of the one who receives the miracle. The recipient's faith was an integral part of the miracle, as when Jesus healed and then dismissed a diseased woman with these words, "Daughter, your faith has made you well; go in peace." (86) In that instance, the woman uncharacteristically asserted herself and pushed through a crowd to touch Jesus' garment. Jesus sensed what she had done and, in concert with her faith, the miracle of healing happened.

In the next chapter, we will move from a discussion of openness to the spirit of wonder and the engagement of miracles to a skepticism that doubts supernatural possibility. This chasm of discernment often causes people of science and people of faith to be wary of each other. I will attempt to bridge some of this skepticism by challenging each discipline.

TRACES OF ARROGANCE IN SCIENCE AND FAITH

For I am persuaded that your love is established forever;
you have set your faithfulness firmly in the heavens.

—Psalm 89:2

Religion and Science obviously represent two different meridians on the sphere of our minds, and it would be wrong not to keep them separate… But these meridians must necessarily meet somewhere at a pole of common vision. Otherwise our whole intellectual and cognitive enterprise collapses. (87)

—Teilhard de Chardin

It is my perception that people of faith who eschew science and people with a strong science bent who look askance at faith do not have the richness of perspective that those who embrace both science and faith experience. There are, of course, many scientists and people of faith who have embraced both disciplines and feel enriched by them. This chapter is addressed to those who insist that the two are primarily antithetical.

I invite those of you in this category to open your minds to the possibility that science and faith are opposite sides of a very thin coin of great value; their qualities rub against each other. To make that case, I have challenged scientists and people of faith with the following thoughts:

Men and women of science don't have to be so wary of faith. Scientists actually demonstrate faith regularly in their daily life. Moreover, it is often a deep faith perspective that draws them into their special field. For example, most scientists believe that knowledge is good and desirable. That is a philosophical statement without an easy proof, particularly because knowledge can enable evil as well as good. Atomic power is just one strong example of science enabling possibilities for good or ill.

Perhaps, as we get further away from the paradise of "Eden," the closer we are to the possibility of self-destruction. The jury is out on that. There is no observable proof either way. Empirical evidence is the foundational sine qua non of modern science. Yet, in faith, scientists continue to do research and make discoveries and trust that their work will benefit mankind.

However, faith can be two-edged. Misplaced faith tends to divide and alienate. Faith in an object, even the Bible, is misplaced faith. For Jews and Christians as well as many others, faith is in God and in God alone. The Bible lives only through the breath of the Spirit. People who reach in faith for God are aided by the text but ought not to be chained to it. The Bible is a reflection of a people's faith and a response to faith, not the proof of it.

One does well to approach the Bible with the spirit of faith not in its inerrancy but in the God who is revealed through the divine/human saga it presents.

The Bible, often called "The Greatest Book on Earth," is a testimony to the journey of a particular people and their consciousness of God. In it and through it, the God of the Hebrews and later the Christians evolves in the minds of the writers of the Scriptures. The Bible is a book of spiritual evolution. The Hebrews and the early Christians were growing in their spirituality, in their understanding of God. Over the centuries, the scriptural writers reflect that growth.

The evolving Scriptures should cause no one problems; they reflect a rich mixture of the old and the new. Biblical historical criticism began in eighteenth century Germany. Julius Wellhausen is often considered the founder but many scholars followed in his footsteps and most scholars today value those mostly German contributions. Although a good deal of the biblical text as we have it today has been edited by early Hebrew scribes, those who finally set the canon in place had the wisdom to allow many early and later writings to exist together. Thus, the primitive and the more sophisticated writings sometimes sit side by side, allowing us to reflect upon the thought of ancient and more historically based storytellers and historians. The Scriptures are not presented to us in perfect chronological order. However, studied with faith, historical analysis, theological acumen, and common sense, the whole body of literature is a testimony to the meeting of the conscience of humanity and the mysterious presence of God.

The Bible's main thrust has little to do with the historical "when" and the biological "how" of Creation. The wonderful opening gambit in Genesis captures the faith that God is the Creator, and that He is still at work in what many now believe to be a continuous creation. The irony of getting tossed out of the mythical and metaphorical garden is that we were forced to grow up spiritually; we humans now had choices. In a sense, we joined God as co-creators and stewards of a portion of Creation. For many, the gateway to life beyond Eden led to a holy enterprise, ongoing and open-ended.

In the opening theme of Creation, faith and science have every opportunity to dance in their complementarities, covering each other's weaknesses and, like a good marriage, becoming stronger as a result. A person of faith trusts that God is eternal, that God's spirit is present and approachable.

Most scientists suggest a Creation that extends in time for an indefinite period. This known universe is presently thought to be thirteen billion (and change) years old. Some cosmologists postulate that ours may be only one universe among several. Some suggest that our universe operates in four or five dimensions. (88) It has even been posited that there may be universes with "nothing" in them. Fancifully, it occurs to me that these "nothing" universes are where God goes on vacations.

In any case, Creation or Creations may extend well beyond thirteen billion years. That staggering time period ought to be compatible with good theology. While we do not yet have the scientific tools to know what other Creations may have been like before the Big Bang, it makes good sense that there were others, especially from the theological view that God is eternal. Both scientific and theological understandings of Creation involve a long process of constant growth and change.

People of faith embrace the conviction that God's "life" transcends the age of this universe. Thirteen billion years seems so paltry, so finite, so unimaginative, so crippling when compared to God's eternal dimension. Of course, the "time" before the creation of the universe is euphemistic as there is no such reality as time without space and matter. Nevertheless, "God's time" is beyond that which we commonly call time within this universe. So, with great license we can still dare to ponder the question, what was God doing before fourteen billion years ago? It would seem to me that His creative energy would still be working.

Back to our own universe and our own time, scientists and people of faith are keenly interested in whether the universe we live in was designed with a purpose. Does the universe itself offer any clues or are we human beings left to figure it all out by ourselves?

Cosmologists and theologians alike weigh in on this question, and both perspectives add to our understanding. Many fields of knowledge from philosophy to anthropology, from psychology to sociology, from cosmology to theology affirm that human beings need to have a sense of purpose to live. And that sense of purpose stems from the underlying idea that the universe itself, the created order, is indeed purposeful, not accidental. The exquisite beauty of the universe, as well as its quiet orderliness, seem to lay the groundwork for lives with choices and purposeful existences. The created order has given us parameters within which to work.

While faith perspectives and science perspectives approach the questions about our possible purposes in different ways, it is helpful when scientists and people of faith respect the other's way of operating and open their minds to the other's insights.

Let us leave this global community for a moment and stretch our imaginations beyond this planet to the larger universe. We know that the universe is unimaginably and exquisitely large. The pictures we acquire from the Hubble Space Telescope fire our curiosity and stimulate our imaginations. How limited it is to assume that life is not abundant and that the Spirit that we encounter here does not exist elsewhere! From a faith perspective, if the universe is bountifully filled with life, why wouldn't this creator God manifest that [illegible] Spirit in many other places?

We may not be able to answer that question in any depth for a long time, if ever. But with faith and science joined in partnership, the human community may inspire the work necessary to obtain more specific answers. Though we are a relatively young species, we have been around long enough to be confident, through our developed faith, that the Divine Spirit would be recognizable and familiar in all places. For example, we would not expect to have the Spirit encouraging love in one place but hatred in another, or espouse justice in one place but not in another. We would recognize the Spirit's attributes anywhere.

What people of faith get right and what they get wrong

What people of faith get right is their appreciation for the mystery of God's presence. The encounter with God and the connections to this mystery are possible in many ways, certainly through meditation or prayer, and equally so in acts of kindness or in the pursuit of justice.

What people of faith sometimes get wrong is their tendency to particularize and compartmentalize God. When some religious leaders attempt to claim that God dictated an inerrant Bible, they misrepresent the essence of faith. Faith by definition trusts in God's presence without the necessity of proof. If God's presence could be proven in the Bible, faith would be unnecessary. Those who are prone to a literalistic reading tend to put the Bible in a restrictive box that shears faith from its pages. The Bible needs to have the wind of the Spirit blow through it. Faith is a dynamic entity. The Bible lives vibrantly in that tension between faith and the various challenges that appear again and again in the Bible's stories.

Two examples: Fear and faith commingle in the epical story of Jacob wrestling all night with the angel, perhaps his own conscience, before meeting his brother Esau in the desert after twenty years of estrangement. (89) Or we can look at the encounter that Jesus had with some rulers of the temple who brought to him a woman caught in adultery when the law allowed them to stone her. The test of faith came when Jesus calmly said to her persecutors, "Let him who is without sin among you, be the first to throw a stone at her." (90)

Biblical Scriptures have occasionally refined the ethics of one era, and in so doing come closer to what we value as divine. The early Hebrews were at first influenced if not bound by the widespread Code of Hammurabi, the Babylonian lawgiver, 1792-1750 B.C. (91) Through the course of Hebrew history, the Hebrew leaders increasingly attempted to balance justice with mercy. As the Hebrew culture became more established, laws such as lex talionis (92) that included "an eye for an eye and tooth for tooth" (93) were reduced to economic sanctions. The moral legitimacy of hating one's enemies gives way in the Book of Proverbs to a very different kind of spirituality. "If your enemies are hungry, give them bread to eat; and if they are thirsty, give them water to drink: For you will heap coals of fire on their heads." (94)

St. Paul paraphrases this idea in Romans 12:20, when he writes, "If your enemies are hungry, feed them; if they are thirsty, give them something to drink; for by doing this you will heap burning coals on their heads." (95)

Jesus, drawing on similar passages and embracing his mystical tradition continues to challenge human tendencies to retaliate against perceived injustice. Instead, he advises us, in the face of those who hurt us, to "turn the other cheek" and "love one's enemies." (96) People of faith may rejoice in the spiritual evolution of the Bible itself. Scientists may want to join them in that celebration.

What men and women of science get right and what they get wrong

What scientists get right is a common open-minded perspective that enables them to reach higher and deeper, to stretch the minds of all humanity, to question theories taken as "gospel." Their perspective allows for growth, the reworking of an idea, and an appreciation of what it means to stand on the shoulders of their predecessors without whom knowledge would be stagnant. What scientists get right, as well, is that they often take reasonable risks for the sake of the "truth" as they understand it. Sometimes they must endure ridicule when a new or refined way of interpreting reality abuts a precious older view. Often they must be people of courage.

Scientists occasionally miss the mark on questions of faith. At times, scientists denigrate faith, citing spiritual perspectives no longer held by serious people of faith. Scientists seldom challenge the ideas and concepts of the most forward religious thinkers, nor do they challenge noble human acts motivated by religious beliefs. Perhaps stemming from our medieval history, where religious politics dictated the boundaries of scientific research (à la Galileo), scientists have had an uneasy and sometimes hostile view of religion and religious faith. One only needs to recall the silencing of Galileo during the Inquisition or his excommunication from the Church because he sided with his predecessor, Copernicus, in affirming the heliocentric hypothesis.

While scientists' work involves testing ideas and theories and gathering empirical evidence, their personal lives may be guided by faith as much as anyone else. At times, to cross a street can be an act of faith. There is no proof that one will get to the other side. When a scientist falls in love, the process is not just about experiments and probabilities; he or she needs a leap of faith to bridge the chasm between what is unknown and a promised commitment. That is similar to the way one moves toward faith itself.

Living life demands faith; much cannot be proven; much must be risked. It is the tension between the unknown and the expected that adds zest and joy to life's journey. Scientists and people of faith alike ought to embrace the idea that to live without hope and faith is foolish.

Love, like faith, cannot be systematically measured. Religious faith at its best is about love and how to share it, not so much about how to get it for oneself. An axiom of religious faith holds that one receives when one gives, not the other way around. Jesus said it in a slightly different way, when he said, "It is more blessed to give than to receive." (97)

Science does not usually venture into this arena, though some from allied professions have explored it, as in the sociology or epistemology or philosophy of love. Scientists presumably need to love as much as non-scientists. Therefore, the best of the field of religion is still open and available to them. Religion is about how to live with one's neighbors and one's enemies, about the passion for justice and mercy. We point to the prophetic wisdom that still finds its way into our daily lives: "Let justice roll down like mighty waters…." (98) Scientists, and all those espousing a rational approach to living, presumably need to live well with their neighbors, and learn to live neighborly with their enemies. Therefore, the field of faith is pertinent to everyone.

In addition, the scalpel of introspection is needed for people of religion and faith as well as those for whom empiricism and a scientific perspective are the only guiding principles. As a Christian, I am occasionally distressed and embarrassed by a large and growing number of Christians who shun science altogether. In too many Christian circles, there is a surrender of the mind to a kind of blind faith; biblical scholarship is suppressed, leading to exclusivity and arrogance. The hallmark of faith should be humility and openness. In too many places, the well-balanced Christian, who loves God as much with his mind as with his heart, is fading into the historical landscape.

Scientists and people of God, it is time to wake up! We need you. We need science with a heart and religion with a brain. As were heralded earlier in the book, some heroes who combined these perspectives are no less than Galilei Galileo, Isaac Newton, Albert Schweitzer, astronomer Maria Mitchell and Albert Einstein. In this transitory world, this is no longer just an academic exercise—the world needs both disciplines now. People of faith, we cannot kill the Spirit by hiding from knowledge. Scientists, we cannot kill the mystery of living by hiding from spiritual power. We are on a precipice, and if we are not more comprehensive in our thinking and acting, each of us will fall into the abyss dragging the other with us. Together, however, we can build a bridge that will span any chasm, and the view from the bridge will be glorious.

In later chapters, I will explore concrete ways in which science, specifically the field of medicine, and faith can play mutually beneficial roles. I will address some of the issues surrounding stem cell research, the environment and space exploration.

A BRIDGE OVER THE TROUBLED WATERS OF PAIN AND SUFFERING

Although the world is full of suffering,
it is also full of the recovering of it. (99)

—Helen Keller

We cannot learn without pain. (100)

—Aristotle

Human beings in every land and from most every spiritual perspective often find it difficult to believe in a personal, loving deity when faced with so much pain and suffering in the world. This apparent incongruity causes a rift in our minds and hearts that creates a spiritual chasm too great for some people to overcome. In this chapter, I will try to construct a bridge over these troubled waters, supported by the common link of suffering that connects all of humanity.

Sometimes the pain and suffering that human beings experience seems to bear no relation to the quality of life people have lived. Pain and suffering are sometimes of our own making, but often, it must be stated boldly, the causes do not stem from our sins or mistakes. Diseases can strike with indiscriminate vitriol; physical catastrophes may destroy cities, independent of the faith and goodness of the populace. Crime can be visited upon people in a random way. Suffering is often not the victim's fault.

It is also well noted that pain is inevitable. Suffering, however, is how we deal with pain, and we do have some control in that area. Our minds and spirits come into play and influence the extent to which we will suffer. While medicines can reduce pain, many faith groups and other meditative groups teach people how to cope with pain and reduce suffering.

Our vulnerabilities invite the question: How does a good God allow so much suffering? Many people disavow an allegiance to God when they have been seriously hurt by circumstances that were beyond their control. The blood of innocent victims cries out to the living.

As a pastor, I have known many people whose faith ebbed or disappeared when confronted by their own pain or by the untimely loss of a loved one. Without knowing it, sometimes we make a bargain with faith. As long as life is manageable, we will have faith, but if it hands us more than we can bear, we may discard it. God, we reason, has let us down.

Literature is full of writing that has attempted to get a theological handle on the causes of pain and suffering, including the bestselling book, *When Bad Things Happen to Good People* by Rabbi Harold Kushner. Rabbi Kushner suggests that God has given us everything we need to have a dynamic and fruitful life, but He does not ever enter the human realm that we call the here and now. We are left on our own albeit with extraordinary God-given capacities. We can call on the Spirit for strength and wisdom, and even the healing energies with which we have been endowed through creation, but we cannot call on the transcendent God to change events. "God has set Himself the limit that He will not intervene to take away our freedom, including our freedom to hurt ourselves and others around us." (101) In this regard, God is off the hook. There is no sense getting mad at Him when bad things happen; He does not interfere with the laws of nature. That, of course, is just one idea on the subject, and seductive as it is, it does not inspire me. As a believer in the Christian understanding of God, I find Kushner's arguments personally unsatisfactory. I am persuaded through the body of Scripture that God stays involved with His creation, His immanence is as important as His transcendence. God has not left us on our own; His presence within and without us is vital yet mysterious, not accessible to much scrutiny. But, adopting this position makes is much more difficult for humans to answer the question, why is there so much suffering if there is a loving God?

Mary Baker Eddy, the founder of Christian Science, held that since God is good, all evil and disease are of our own making, an error in our thinking that infects our bodies. (102) Some evangelicals believe that common disasters and catastrophes are indications of the wrath of God, as in some biblical interpretations of the Great Flood. I heard a well-known TV evangelist describe the tragic 2010 earthquake in Haiti in those terms. This may be the weakest of all interpretations leaving little room for compassion for innocent victims. It is too easy an argument and too narrow-minded in scope.

Other interpretations of pain and suffering invite more thought. Carl Jung, the great Swiss psychologist, wrote, "Suffering and conflict are a part of life; they must not be regarded as 'ailments'; they are the natural attributes of all human existence, the normal counterpole, so to speak, of happiness." (103)

Some religious interpretations offer the palliative that, though the travails in this life may be great, if one goes through them with humility and faith, great rewards await in the next life. This point of view emphasizes that this earthly life is a preparation for the next and that suffering is a key test of one's character and faith. The Gospels themselves often suggest that eternal rewards will be merited for those who live by blessed standards as expressed in the Sermon on the Mount.

> "Blessed are the poor in spirit for theirs in the kingdom of heaven...Blessed are you when people revile you and persecute you and utter all kinds of evil against you falsely on my account. Rejoice and be glad, for your reward is great in heaven." (104) As a Christian I take seriously this point of view, but I also recognize that this earthly life is the beginning of an eternal walk, and suffering is more than preparation for the next life. Suffering humbles and shapes us and can make us more empathetic to suffering of others. It has this-worldly purposes as well.

To some who look at the question of God's role in pain and suffering with a dispassionate and perhaps scientific view, it is the natural laws of physics that make the most sense and reveal the most important reality. They would suggest that one must look at the big picture. First, they remind us that the planet's environment must be refreshed from time to time to continue the planet's viability for life, and second, populations need to be controlled if the planet is to be a vibrant host to many species. Natural catastrophes, earthquakes, volcanoes, comets from outer space, and violent storms and fire allow this planet to rejuvenate and retain its vitality. The short-term violence has always been the price of an otherwise magnificent Earth. If one were to put an ethical spin on it, one could argue that victims of natural disasters are the sacrificial lambs for the rest of creation.

From a human point of view, it is almost impossible to conceive of a world without pain and suffering. These conditions are built into our very nature as well as into the physical realities of Earth. Philosophically, we have learned to adjust as well. Without pain and attendant suffering, we would not know joy and the happiness that follows. The two seem to have meaning in relation to each other. Pain and suffering are part of the essentials of a dynamic, evolving life. Without suffering and the quest for happiness, life would be static, and motivation would yield to inertia.

Men and women of science are among those who have actively responded to human pain and suffering. The field of medicine was created by people motivated to do something about disease, believing that with medical knowledge and institutional collaboration we could more effectively ease human suffering. Medical professionals don't necessarily answer the question of why disease happens, but there is no doubt that their actions improve the quality of human life to an extraordinary degree. Through their medical arts they heal untold millions and alleviate many of the world's woes.

Men and women of faith have a different way of responding to humanity's suffering than the medical profession. The mission of people of faith is to care, not necessarily to cure. Their goals are complementary to those in professional medicine. Most religious institutions attempt to respond practically to human suffering. They too may not answer the great questions that lie behind suffering, but the testimony of their faith is in the quality of their care.

One of the best examples of faith and care moving into action that responds to human suffering is the world-wide organization called Special Olympics. Born in the mind and heart of Eunice Kennedy, at first a day camp for children with mental disabilities in the early 1960's in Maryland, the idea of giving these children a sense of competition has now grown to a hugely successful international organization. I would like to suggest that the spirit and philosophy of Special Olympics alters the normal understanding of evolution to which we have grown accustomed. Survival of the fittest, axiomatic in biological circles is thwarted in order that a higher standard of human evolution may exist. Bigger, faster, stronger, smarter, craftier, goals with which we are familiar around the globe are superseded by a belief that it is possible to put skill, courage, sharing, joy and, above all, compassion above those usual goals. Former NFL back and TV analyst, Frank Gifford writes in the foreword of the book, *Skill, Courage, Sharing, Joy—The Stories of Special Olympics*, "Beyond the gold, silver and bronze medals won by competing Special Olympics athletes, beyond the ribbons, hugs and cheers, there's a higher prize to be won…a prize for the way the game is played." (105) Examples of this ethic abound at S.O. such as when a particular young girl was winning a race. "Over her shoulder she saw that a friend had fallen. She went back to help her friend rise, and hand in hand they finished the race—in last place. But in the hearts of most who observed them, they were first." (106) That image now adorns a Special Olympics medal.

In the summer of 2011, I had the privilege of giving awards to tennis players at the Special Olympic World Games in Athens. Giving ribbons on a podium, not only to first place finishers but to fourth and fifth place finishers testified to this higher value. Participation and the journey itself trumped the idea that winning is the only thing.

President John F. Kennedy, Pope John Paul II and even Gandhi, along with many other notable speakers, are reputed to have said that civilization's greatness can be measured by the way it treats its weakest citizens. Human evolution has come to a point where some are not willing to set aside those who are not "fit" but rather change the definition of what is fit. Love and care become the means and the ends in this arena, and a new world is born. It reminds me just a little of words that Bishop Esquirol of Connecticut preached at my installation as a rector of Trinity Church, Collinsville, Connecticut in 1969 when I was a twenty-nine year old fledgling priest. He said, "God does not call those who are fit; He makes fit those whom He has called." Those inspiring words have remained in my mind through the years.

Other communities of faith have dedicated themselves to regular prayer for the victims of disease or social difficulties, the kind of prayer that leads to action. The power of community prayer is a key to the strength of many faith institutions. The very action of prayer often motivates people to care more for the victims, to visit them, to offer financial help, to help build hospitals or shelters for the homeless, to be present in one way or another. It is important to remember that while it may be difficult to demonstrate the effectiveness of individual prayers for the healing of someone who is ill, it can be demonstrated that "the work" of prayer of a community of faith buoys and gives vitality to the community itself. For example, someone who did not know of the travail of another hears through community prayer the difficulty of another person and immediately visits them or makes a meal or performs some other thoughtful service. These actions provide solace to the one who is ill or hurt, the feeling of accomplishment for the one who gives the service, and the totality of these experiences strengthens the community of faith. Some would say that we become the hands of the Divine Healer in so doing.

During the 2010 flood that caused so much damage and heartache in Nashville, Tennessee, I heard from a former high school classmate who just buried four of his friends. He sent me some notes relayed by his pastor at Nashville's St. George's

Church, reflections from some of his congregation. They serve as a testimony to caring people lifted up by the power of a praying, caring community. "I have never prayed harder for everyone at St. George's and for God's peace. God has heard and answered my prayers." "God is enabling me to sleep at night despite the losses." "We are so up lifted by all the prayers." "It's incredible to see the kindnesses extended by so many—Complete strangers helping one another." "I've met parishioners I never knew before. We are working side by side at the church." And the pastor responded, "Please take a moment, pause, and really listen to each other. We are the eyes, ears, hands, and feet of Christ. Don't miss this opportunity to minister to one another." (107)

Outside faith-based communities as well as within them, some institutions exist for the primary purpose of addressing the needs of those who suffer. There are social scientists who treat mental anguish and mental illness. The causes of these problems are multifaceted but, once again, human beings with various capacities and ideas about treatments are motivated to care and to cure. Compassion combines with knowledge and skill to affect healing.

Overcoming or managing addiction involves both the skill of social scientists and the faith of the person who suffers from addiction. Treating those with alcoholism links faith and science. Medical doctors may prescribe drugs to enable a patient to begin a social process of recovery. The twelve-step program of Alcoholics Anonymous is perhaps the most successful of all programs in helping alcoholics recover. It is so successful that programs for other addictions, such as gambling, sex or other uncontrolled compulsive behaviors, have borrowed the principles of the twelve-step program.

The first two steps are particularly germane to our discussion. Step one is the admission by the person afflicted by this addiction that he does not possess the power or ability, by himself, to control his drinking. Step two is the recognition of a higher "power" greater than any power the person possesses, that can restore the addict to health. (108)

In addition to institutions that exist for the primary purpose of easing human pain and suffering, many other institutions also help, though involvement may be subtler or more indirect. Ordinary insurance companies came into existence to help in times of crisis and to ease suffering. Likewise, our social security system, health care plans and even banks exist, in part, to ease economic suffering and enhance health and prosperity. The entertainment industry, while it often aims to divert us from our immediate concerns, can also make us aware of larger social problems and enlighten us about how we may be part of the solution. Contemporary examples include movies such as Invictus, a story of Nelson Mandela and issues of apartheid in South Africa; Slumdog Millionaire, and the awareness it brought of poverty issues in Mumbai, India; and Precious, and its exposé of the trauma of childhood abuse. It is also apparent that many individuals who are not people of active faith also aid those who suffer. They would probably give as their reason for doing so that they feel the drive to help as part of their human nature, not inspired by an outside force, perhaps part of the evolution of human beings themselves. When this happens people of faith and people of "goodness" without faith join together. Perhaps, each can appreciate the other, as happens regularly in hospital institutions.

Most people find the causes of pain and suffering to be, in part, mysterious. Although our society has different ideas about the role of faith in addressing inevitable sorrows, people throughout the world take an active role in institutions whose purpose is to benefit humanity and to ease pain and suffering. Their noble actions are the greatest testimony of faith in the created order. No philosophical or theological system has completely answered the problem of pain and suffering. But the active human response to suffering on the part of people of faith is a strong testimony to their belief in the power of the Spirit in the world

A BRIDGE FOR MEDICINE AND FAITH

Daughter, your faith has made you well; Go in peace.

—Luke 8:48

I wanted to be a doctor that I might
be able to work without having to talk. (109)

—Albert Schweitzer

For thousands of years, medicine and faith have enjoyed a symbiotic relationship. In the distant past, medicine was not as sophisticated as it is today, and faith and medicine shared a rather equal import. In this chapter, I would like to explore the relationship between faith and medicine and strip away any modern tension between the two. First, let us look a special aspect of the ministry of Jesus. The Galilean teacher was also a healer, and he often wove into the healing relationship the statement, "Your faith has made you well." (110)

In the Gospel of Luke, a woman who has been suffering from hemorrhaging for many years approaches Jesus in a crowd of people. She boldly touches the hem of his garment, and Jesus turns and demands to know who touched him. Because the press of the crowd was strong, the question embarrassed some of the disciples. But, Jesus would not be deterred. "I felt power leave me," he said. The woman confesses that it was she. Jesus responds to her, "Daughter, your faith has made you well; go in peace." (111) If judged by the casual contact accepted between men and women today, the incident may seem inconsequential. But, in the Palestinian culture of Jesus' time, a woman was not permitted to touch a man in public, and to touch a teacher was a strong taboo. It required courage to do so.

Jesus was hardly the first to suggest a relationship between faith and healing. Few would deny that the disposition of one's mind and spirit is important for optimal healing. Even further back, in the eighth century B.C., the prophet Elisha, known for several remarkable miracles, healed Naaman, the commander of the Syrian army, of the dreaded leprosy. The healing required that Naaman act in faith. He was initially angry at the simplicity of Elisha's request, but humbled himself and went and washed seven times in the Jordan River. The response in faith cured him. (112)

These biblical examples of healing suggest that an important ingredient to curing all manner of illnesses is faith. Jesus brought the element of faith into most of his healings. For the hemorrhaging woman, one could wonder how much was due to her faith and how much was due to the healing power of Jesus. In the story, Jesus relates that he felt a power go out from him, but concludes that the woman's faith had saved her. The two are not mutually exclusive. But, as I will illustrate in this chapter, an all-or-nothing approach to faith may, at times, be dangerous.

Medicine is, among other things, the science of diagnosing, treating or preventing disease. What formerly was the purview of faith practitioners is now predominantly in the hands of doctors and the medical community. The relationship between faith and medicine is, perhaps, less secure than it was.

As a priest, I have visited many patients over the years. Upon entering a hospital room, it was obvious to me that medicine and faith were complementary to one another though practitioners of faith are sometimes regarded as the poorer sisters of doctors of medicine. Many patients who rely heavily on the medical profession are thankful to have a priest or minister or rabbi present to buoy their spirits and pray for healing. Some would say healing the spirit is as important as healing the body. Hospitals have melded well the doctor of science and the spiritual guide. Each has its place.

In this chapter, I explore in greater detail the interrelationship of faith and medicine. I will include some real-life stories of people with sickness or injury. Sometimes faith and medicine work in concert, and at other times the two mix in a confusing way. That is the case with this first story.

In 1954, when I was in the eighth grade, a young girl in my class named Patty, who happened to be an identical twin, became sick. No one knew the cause or the prognosis. She was a Christian Scientist and her family did not take her to a doctor. She was out of school for about eight weeks and, because of the family's faith, the community did not know how sick she was. Her illness was shrouded in secrecy except, I suppose, to a very few. Patty never came back to school. She died. But we learned later that her family had rushed her to the hospital a few hours before she died. By then it was too late. It was rumored that she died of leukemia. If that is so, then perhaps, at that period of history, she could not have been saved anyway. We will never know for sure.

The community was stunned. We whispered about the issues of medicine and faith. Some were very angry; all were heartbroken. The more we talked, the more we began to understand the tension that is possible between relying solely on faith or solely on medicine. Even young teenagers, like myself, sharpened our grasp of the issues involved. The poignancy of this issue was additionally brought home by the fact that Patty's family later changed the expression of their faith to one that incorporated medicine and faith.

In talking about this tragic event at home, my mother related to me that when she was in college she became a Christian Scientist and was earnest in studying the faith for a couple of years. She then had an acute appendicitis attack and had to choose whether to see a doctor or to rely solely on her newfound faith. She picked a doctor.

Interestingly, my mother never relinquished part of her Christian Science training. She always believed that complaining about one's health usually made one sicker, and that a positive attitude was necessary to be healthy. She made a good compromise and integrated well the two disciplines of medicine and faith. Ironically, she was a serious smoker for sixty-five years. In all probability, it finally killed her, but not until she was ready to go. She was well into her eighties.

Sometimes the relationship between medicine and science appears quite literally at the last trump. In the late 1970s, when I was the rector of a small church in Collinsville, Connecticut, I visited a man who was gravely ill and in the hospital. He was not a churchgoer and I had never met him, but he was on our parishioner rolls. As I walked down the corridor, I heard the beeping from the monitor at the entrance to his room change to an alarm. It displayed a straight line going through the middle of it, a sign that the patient's heart had stopped. As I came to his room, nurses were running; a trauma team arrived within a minute or two. They immediately set to work to save him and successfully shocked him back to life. The second act of this man's life really begins here.

A bit of background: This gentleman was unusual. It would be kind to say that he was a curmudgeon, more accurate to say that he was a crotchety old man. His very large family lived in his area. None of them came to visit. He had at one time or another offended them all.

The nurses and doctors knew none of this. They knew that they had just saved a man. He was their Lazarus. He became precious, especially to the nurses who took care of him around the clock. For a long time he could not speak, so the nurses applied their healing ministrations amid his silence. Slowly but surely he grew stronger.

I continued to visit. This latter-born Lazarus interested me as well. On one occasion, he was lying back and his right hand was resting on a clipboard. When I gently touched his forehead, he began to write without being able to see what he wrote, so the writing was very scratchy. Nevertheless, I was able to make out that he said: "Thank you for coming." There was a faint smile on his lips. These were the first words he had communicated to me. I felt a lump in my throat.

Many weeks later he left the hospital, but the nurses had a hard time letting him go. When a patient leaves the hospital, that exit is often the end of contact for the hospital floor nurses. Instead, several nurses visited him at home, though it was forty minutes away from the hospital. They came initially out of duty, but later just because they wanted to. He was their hero; they were his heroines. They saw in him not the crotchety old man that others had known but a man of hope and appreciation and tenderness. It was a metamorphosis. This human saga was not about pure science; it was the best of science and the best of relationship born of the active faith of a small medical community.

An important part of the healing process involves the will of a patient to be healed, and in many cases a person must actively participate in his own restoration. A classic New Testament story illustrates this point. In ancient Jerusalem there was a curative pool called Bethzatha, also called Bethesda. It was locally believed that the first one in the pool in the morning could experience its healing powers. The Gospel of John tells the story of a man who lay on a stretcher and was placed by the pool every day hoping to be put in first. He had been taken there regularly for thirty-eight years. On one occasion, Jesus was present and confronted him with what seemed initially like an insensitive question. "Do you want to be made well?" (113) Jesus sensed that the man was resistant to healing. The man blamed his ongoing condition on the fact that, in all that time, no one had put him in the healing pool. Sometimes we lack the courage to dare to be well. One must participate in one's own healing.

During the forty years of my active ministry, I witnessed a great blossoming of a rite known as "the laying-on of hands." Many religious communities share some common ground and faith in this physical/spiritual endeavor. The rite can be quiet and subtle or loud and ostentatious. On the low-key end, the Catholic Church and the Episcopal Church have developed liturgical forms that incorporate healing rites. The Episcopal Book of Common Prayer, with which I am most familiar, presents such a service. (114) The service is often a quiet, weekday service without any grand display. In our tradition, the priest officiating at this service of healing prayers lays his hands on the head of the parishioners saying words such as these:

> "I lay my hands upon you in the name of God beseeching our Lord Jesus Christ to sustain you with his presence, to drive away all sickness of body and spirit, and to give you that victory of life and peace which will enable you to serve him both now and evermore." (115)

Evangelical Baptist churches often have more elaborate ceremonies, in which a person or persons to be healed come up on the stage, perhaps witnessed by hundreds of parishioners, and the healing minister is quite dramatic in his ministrations, denouncing the devil and calling the Lord of Heaven to grant physical and spiritual healing to the recipient.

Other traditions have their own rites, expressing similar sentiments. Some have an entirely free form nurtured by the power of the spirit present at that moment. All faith groups affirm that healing happens in one way or another. Perhaps there is no dramatic physical healing. Instead, the action may be a way of centering a person; the prayers and physical touching may foster a spiritual focusing upon personal matters. Laying hands on a person's head or shoulders may reduce anxiety enough so that a person can see a better path more clearly, hopefully a path directed by or in harmony with the Spirit of God.

Services that include the laying-on of hands have three elements. First, the community of faith offers the service girded in biblical tradition. Second, the pastor or layperson who lays his hands upon another person brings his own faith to the engagement. Third, the person who receives the laying-on of hands brings his own faith to this moment. We do not try to figure out which is more important. What we experience is that when all three energies are present, healing will take place, not the healing of our designs, but that which God dispenses. We do not dictate to the Spirit of God what that will be or how it will take place. We place ourselves at the Spirit's disposal and let go of our need to control. We walk out in faith.

In Sarasota, Florida, a retired Jewish friend of mine, who had formerly been president of his synagogue for many years, introduced me to a community called Metaphysical Christianity. (116) He invited me to his church and prepared me a little for the experience. Metaphysical Christianity heralds Jesus as a mystical healer, but its followers do not regard him as the Son of God. When one enters the church, one is drawn immediately to a painting that covers the front wall of the sanctuary, a twelve-foot painting of Jesus in the act of healing. For those of this tradition, Jesus is the model healer, but his powers are equally available to many who pursue the healing arts.

The Sunday I attended was a typical service. At the beginning, the minister welcomed the congregation, and then about fifteen leaders, trained in healing and garbed in white vestments, came to the front of the nave and stood behind fifteen or so vacant chairs. The minister invited anyone who would like to come forward to sit in the chairs. These men and women would then barely touch the heads and shoulders of the people who sat before them. This lasted about five to seven minutes. I came forward, sat in one of the chairs and experienced the healing prayer and touch of the person behind that seat, and also the communal strength of the congregation at prayer. I surmised that if one did this regularly, much like the healing services in my own tradition, that over time, healings of various kinds could be experienced. The corporate strength of the community sharing the healing prayer could not be dismissed.

The healing touch of traditional doctors is of more value than is sometimes perceived. A surgeon who visits a patient before the operation and shakes his or her hand or places a gentle hand upon the shoulder can have a remarkably positive effect upon the patient. I still remember a doctor's visit to my hospital room some forty years ago on the afternoon before I was to undergo surgery. His warm handshake enveloped my hand, and my spirit soared. Just that touch filled me with confidence, and my fear dramatically abated. In the hearts and minds of patients, many people in the healing professions carry with them an almost mystical presence, which in itself can be powerful medicine.

I know a man who was just about to have back surgery at a large New York hospital where several patients were wheeled into a surgical waiting room some minutes before their respective operations. In the line-up of gurneys, this man's head was up against another patient who was also awaiting her operation. The man heard her sob and, reaching his hand backward over his head, he took her hand and held it for a couple of minutes. A week later, a nurse came to his room with the patient in a wheel chair. The nurse said, "This woman wanted to meet the person who held her hand just before surgery. She just wanted to say thank you." Such is the power of human touch.

Of course, the relationship between modern medicine and traditional faith is complicated. The two fields have their areas of "expertise." Some say that the mission of a medical doctor is to cure, while the mission of a minister or one in the healing arts is to care. Traditional ministers who learn this subtle difference can greatly reduce their own anxieties about their role with their parishioners or clients. Both goals are worthy and necessary. The two disciplines also overlap at times. Patients will place their expectations in the hands of doctors and their hopes in the hands of ministers. Many will seek both, and as long as the two fields are respectful of each other, they can exist harmoniously.

Albert Schweitzer, an extraordinary man of the early twentieth century, combined the world of theological faith and the world of medicine as completely as anyone who has ever attempted to integrate the two disciplines. After years of writing and lecturing in august university settings at Strasbourg and later Oxford and Cambridge and preaching regularly, he opted to get a medical degree. Upon becoming a medical doctor, he determined to go to equatorial Africa in 1913, as the only doctor at the mission station hospital at Lambarene. It was a most primitive and difficult setting. In that long-ago day, he was referred to as "a jungle doctor" back in England. (117)

In *Out of My Life and Thought*, Schweitzer wrote about his decision to leave civilization and his friends behind to work as a doctor in this very poor and primitive place:

> "I wanted to be a doctor that I might be able to work without having to talk. For years I had been giving myself out in words and it was with joy that I had followed the calling of theological teacher and of preacher. But, this new form of activity I could not represent to myself as talking about the religion of love, but only as actually putting it into practice. Medical knowledge made it possible for me to carry out my intention in the best and most complete way, wherever the path of service might lead me." (118)

It is easy to understand that Schweitzer, who had such a noble and brilliant academic career, was a hero to many; yet he believed himself to be the servant of Jesus. He did not want to ennoble himself, but rather the One to whom he had dedicated his life. Even the music that he played for the enjoyment of others had an egoistic chord to it. He eschewed that tendency in himself. He yearned to be part of the action of love, not just to lecture about it, or even to entertain, and he found a magnificently humble way to do so.

I am attracted to Dr. Schweitzer's life because it represents a challenge to everyone who is called to serve in either medicine or the traditional ministry. Of his two life choices, one appears to be a mainly scholarly discipline, the other a mainly spiritual discipline. To be whole and integrated, each discipline must overlap with the other. The medical doctor must reach out and touch his patients. The theologian or minister must find an active way to express the theology that motivates him. The doctor must find a way to care and the minister must ground himself in this world. Each can borrow from the other. When they do, their lives become richer and more fulfilling. More importantly, when these professionals work in harmony, the patient or parishioner is so much better served.

BRIDGES FOR LIFE'S ISSUES

I build bridges from one truth to another along my life's way, either making room for the vision or transcending the currents flowing beneath. (119)

—Jacques Linder

Ethical axioms are found and tested not very differently from the axioms of science. Truth is what stands the test of experience. (120)

—Albert Einstein

A healthy dialogue integrating the social consciousness of people of faith and people of science can make a positive impact on many of the social problems that we address today. Three social problems may serve to illustrate this potential—capital punishment, stem cell research, and what I call an "Armageddon mentality." I will examine how our understanding of science and faith aids or inhibits our judgment.

As we attempt to solve these challenging social issues, our attitudes, philosophies, values, and ethics equip us to differentiate right from wrong or at least enable us to negotiate our way along thorny paths and to make choices that will have consequences for our generation and future generations as well.

Over the millennia throughout western civilization, the Bible has had an incalculable impact in the formation of society's ethical standards beginning with the Ten Commandments. It is appropriate to ask, do the ethics promoted in the Bible still inform our decisions? Does science correctly make us challenge old assumptions and offer alternative ways to make ethical decisions? Do science's offerings and biblical ethics help us to make more Solomon-like judgments?

If we turn to the Bible to help form our opinions, then we must be clear about how we interpret the Scriptures. Can a specific decision or judgment rendered in the Bible be strictly applied to a contemporary ethical issue? I believe the answer is, sometimes. I would rather have us look at the whole body of biblical literature and ask, How has this issue been developed throughout the Scriptures? Is it enough to inform our current ethical decision making?

I think the Bible and its development of ethical laws to live by may be more useful as a spiritual and ethical guide than as a binding, juridical document. From a theological point of view, God urges us to stretch our minds and embolden our spirits with the loftiest ethics that the Bible advances. With the aid of our individual, God-given conscience, and with the collective conscience of others, we are encouraged by the "divine spark" within us to reach the godliest ethical decisions possible. If we use the Bible in this manner, each generation of humanity has the opportunity to be engaged in it, to be part of the on-going process of ethical discernment.

Many modern social challenges require the work of our collective conscience, that is, the moral discernment of a community or nation or the world. Let us look first at the issues surrounding capital punishment. Our views on this subject depend in part on the cultural and political climate. The majority view in the United States has flip-flopped more than once in my lifetime. Both science and faith have a role to play. Faith gives us a moral context. Science gives us the data and the tools within a culture to refine our thinking on the subject. For example, each state adjudicates the issue of capital punishment differently. Though people's opinions vary, several questions have recurred through the years. For example, are we absolutely sure we have the right person? Does society seek revenge or justice? Does capital punishment have biblical precedence and authority? Is capital punishment an economically wise choice? Does capital punishment foreclose the possibility of a greater good through the redemption of the convicted individual? Finally, what does it do to the soul of society to be in the business of executing human beings?

With these questions in mind, let us look at what the Bible offers us. We will see also that the Bible has evolved in the direction of mercy as it develops. The issue of capital punishment itself actually predates the Bible. "Lex Taliones," the ancient law of equal retaliation (whose origins predate ancient Hebrew society), is one standard of judgment from the Torah, as seen in Genesis, Exodus, and Deuteronomy. (121) In Leviticus, we read "Anyone who kills a human shall be put to death." (122) Many people advocate this position today. But from another angle, the clarion call of Deuteronomy arrests our attention: "Vengeance is mine and recompense." (123)

That vengeance belongs to the Lord is repeated again in Paul's sublime New Testament Letter to the Romans. He writes: "Beloved, never avenge yourselves, but leave room for the wrath of God; for it is written, Vengeance is mine, I will repay, says the Lord." Paul concludes with this powerful admonition. "Do not be overcome by evil, but overcome evil with good" (124)

In order to use the Bible to interpret ethical issues today, one must be familiar with the ethical evolution that is evident in it. Furthermore, the Bible is set in the context of a much earlier culture than ours. Whether a "capital crime" happened in biblical times or in our own time, the context determines how an individual, community or nation responds. When a "capital" offense occurs today, such as the murder of a doctor who performs abortions, the temptation is to respond with one's own bias and, perhaps, to search the Scriptures to confirm or validate one's own point of view. The Bible is quoted, often out of context, often not from a later text that has matured over time. This response is reckless and too often misrepresents the way the whole Bible deals with the subject.

I suggest that, first, it is more instructive, more helpful, and promotes a better justice to digest the various points of view in the Scriptures. Then we engage the use of our individual consciences, along with the collective conscience of the community to determine how we respond to current ethical dilemmas. I believe God calls us to act with a prayerful discernment, using our full mental and spiritual capacity. The Bible should instruct and influence us, but it is not a rigid law book to be used without our judgment and mature thinking. As usual, the Lord leads us but does not do the ultimate work for us. The door remains open for us to develop and mature.

How can we apply this ethic to the matter of capital punishment? I am instructed by the core of biblical ethics, and I greatly respect cultural values as well. Together, they lead me to be cautious in my thinking and judgment. Most of us are aware by now that, in the United States, keeping an offender on death row awaiting execution is actually more expensive than incarcerating him for forty years. I am also moved by Paul's words concerning vengeance and more especially by his powerful belief that evil is best overcome by goodness. I am influenced as well by the imperfections in our judicial system as well as our human nature. Those concerns lead me to question whether we know with certainty that the right person is facing the executioner's hand. Many people support capital punishment, believing that they are recognizing a need of the victim's family for closure. They think that executing the perpetrator enables the remaining family to begin anew, justice having been served. But, time after time, families in that position recognize later that the execution did not bring them peace of mind or closure. The satisfaction and finality they were looking for and need is not found in executing anyone.

Theologically, final closure is only achieved by letting go of the need for vengeance, minding the aphorism "let go and let God" and ultimately forgiving. Pope John Paul II made a great symbol of this by visiting his would-be assassin in prison and forgiving him in 1981. The penalty still had to be served, but the Pope was spiritually free. Finally, the Bible moves us toward compassion and the chance of repentance and redemption for the one who commits the crime. The time in prison may enable that to happen. Therefore, taking into account all of these matters, I would rarely opt to execute a person.

The second example of a social challenge is stem cell research, a topic that tugs at the ethics of people of faith as well as scientists. Much of the argument against proceeding with research has come from religious people, who cite biblical passages to augment their view, particularly "Thou shall do no murder," from the Ten Commandments. (125) Most every civilized person and government accedes to this ethical demand.

Many religious people have attempted to police stem cell research, about which the general public is not always well informed. At this writing, in the spring of 2010, President Obama has just removed the federal ban on embryonic research. He was greeted with a flurry of criticism as well as support. For example, The Roman Catholic Cardinal Justin Rigali, chairman of the U.S. Conference of Catholic Bishops' Committee of Pro Life Activists, was dismayed by the decision. He responded by claiming that this action would open up the door for inappropriate use of embryos. Simultaneously, the Union of Orthodox Jewish Congregations issued a statement in support of lifting the ban, suggesting that many people would survive and live better lives than they would have as a result of this action. Likewise, the United Presbyterian Church and the Episcopal Church supported lifting the federal ban on embryonic research.

In a democracy, opinions of the people are important since they vote their opinions. So people must be well informed. As we know, "The road to perdition is paved with good intentions." (126) Otherwise we end up comparing apples and oranges, as has occurred in the debate about stem cell research.

We need to step back and ask, are the stem cells used in the research taken from viable fetuses that would grow into human life, and in so doing damage or curtail that life? In my view, that procedure should be unlawful. But, consider a case in which a couple has been successful with an in vitro fertilization, and the remaining incubated eggs are normally thrown away. Why should the cells not be made available for research that could lead to saving other lives? Yes, it is immoral to substitute one life for another, but in many situations the circumstances are not so clear-cut. Some opponents of stem cell research are so unyielding in their attitudes that even when the research and the procedures do not include killing a life or in any way hurting a potential life, they rail against it. Sometimes, however, the dilemma is made clearer when someone in one's own family could potentially benefit from the research. The drama and the heartfelt issues involved were sensitively portrayed in Jodi Picoult's book, *My Sister's Keeper*, (127) and the movie that followed. The story involved a little girl giving some body parts to enable her older sister to live that begged the question: who has authority in this case, the young girl, who was surrendering her "non-essential" body parts, or her parents or the larger society? Tangentially, where do the responsibilities of love and family begin and end?

We see an evolving recognition of stem cell research taking hold in other countries. I have a good friend who twice has traveled to India for six weeks at a stretch to receive daily injections of stem cells. Each of his sojourns has produced remarkable improvement. He has a form of Lou Gehrig's disease, and the advanced treatment was not available in this country when he needed it. Fortunately, he had the financial wherewithal to afford these extravagant trips, but most Americans could not avail themselves of this expensive treatment.

Science can produce amazing findings, not all of which ought to be injected immediately into our society. All of us have a role in being society's ethical watchdogs—using our collective conscience, we are responsible for formulating laws. We want science to inform us about the possibilities inherent in their research, the associated risks, and prognoses. But we don't ask them to decide how society will use scientific discoveries. The best religious ethics, the most spiritually evolved ethics, should play a strong role in those decisions.

A third example of a social challenge that people of science and faith weigh in on is the "Armageddon mentality," that is, the belief that in the near future, the world as we know it will come to an end. It is brought to our attention in the last book of the Bible, "The Revelation to John," a book open to various interpretations. There is no city named Armageddon. The word is derived from Hebrew har megiddon, the hill of Megiddo, where many an ancient Hebrew battle was fought. (128)

Some think that the events foretold in Revelation have already come to pass; others think they portend the future. Some have tried to predict the time when Armageddon will occur, typically the end or the beginning of a century or millennium. Traditional theological scholars think that the idea of an Armageddon that can be predicted is the result of misguided biblical interpretation. One of the consequences of interpreting the Bible loosely is evident today in some sects of Christianity that seem to welcome the planet's demise. When confronted with these doomsday predictions, some of which point to a particular date in the near future, I attempt to point out Jesus' words or the conviction of the early Church on the subject. When Jesus was asked about when the end would occur, this is how he responded: "Heaven and earth will pass away…but about that day or hour, no one knows, neither the angels in heaven, nor the Son, but only the Father." (129) Yet, some modern-day followers of Christ gloss over his very clear statement. More startling, some choose to believe that while Jesus stated that no one knows the day or hour of this cataclysmic event, he didn't mention the month or the year, so it is perfectly okay to try to figure that out. Some who identify a specific date when the world will end believe passionately that when this happens, the righteous will be gathered into Christ's everlasting fold. Further, they believe that Jesus in his "Second Coming" will then reign upon Earth for a thousand years.

Periodically, fervor is created by a new notion that Armageddon is upon us. The recent book *American Priestess*, by Jane Fletcher Geniesse, tells the story of the late-nineteenth-century cultist Anna Spafford. Spafford, a Norwegian immigrant to Minnesota and Chicago, persuaded a large group of mostly Norwegian people from the area, called the "Overcomers," to journey to Israel. There they would prepare for and await Armageddon, when the soldiers of Christ would defeat evil, and the saints would be lifted in what is called the "Rapture." She believed the conflagration would begin in Jerusalem and would happen at the beginning of the twentieth century. (130)

As long as Spafford lived, the community hung together. Her community became known as the American Colony in Jerusalem. It was very strict, and the group concentrated in daily prayer and discipline on getting ready for the end. When the century mark passed, and the end did not come and World War I impacted them, they transformed their quarters into a hospital and ministered to fallen soldiers. Some were beginning to express doubts about the end. When Spafford died in 1923 at eighty-one, the community rapidly lost faith that Armageddon would happen soon and they disbanded. Today, the main building of Spafford's original colony is the thriving American Colony Hotel. (131)

The Armageddon mentality originated from incomplete biblical exegesis and interpretation. This mentality is still alive and thriving. It is dangerous because it invites and nurtures an attitude that the principal motivation for living our lives today is to prepare for the end, rather than to join the worldwide human challenge to improve life on Earth and to be good stewards of the planet.

As strongly as I feel about the mistaken interpretations that can lead to an end-of-the-world posture, I must be careful to avoid arrogance on the other side of this debate. It may well be that the earth will come to an end before its astronomically allotted time. Scientists, particularly astronomers and astrophysicists, remind us that the solar system still contains the potential for asteroids that could severely damage the earth. We also know through normal science that the earth itself will not exist forever. But, scientists measure or predict the end of the earth in millions if not billions of years. An asteroid of any appreciable size only strikes the earth at periods of time measured in tens of thousands of years. Attempts to use the Bible to find a prescribed date for the end are folly. Most scientists and people of faith agree that the time of the earth's demise cannot be predicted. They agree, as well, that it is more productive to direct one's energies to enhancing the lives of people on this earth.

While the history of this planet should warn us about our feelings of permanence, it can also affirm us. Cataclysms are few and far between. Six partial or devastating extinctions have happened in Earth's past. The most recent extinction occurred some 65 million years ago when an asteroid hit in the Yucatán Basin on the shoreline of eastern Mexico, and some sixty percent of life on earth was extinguished, including most of the dinosaurs. In the book Evolution by Carl Zimmer, there is an aerial photo of the Yucatán impact crater that remains as visual evidence of this event. The smoking gun that ties this particular crater to the mass extinction is the level of the element iridium that begins at the crater site and makes its way around the world and is dated to 65 million years ago. (132)

Much earlier, some two hundred and fifty million years ago, a titanic volcano erupted. Scientists believe that ninety percent of life on the planet was extinguished in that consuming cataclysm. These are humbling events for our human psyche to process, but since those disasters have been separated by millions of years, the chances are remote that we will experience one of them.

The extinctions of the distant past may serve as a wake-up call, as the next great extinction could conceivably come from human action rather than an asteroid or volcano. It may come from our own human violence, abuse of the planet due to greed, self-centeredness, and ignorance. Preserving life on the planet may seem a Herculean task, but we are evolving to meet this challenge. It is, perhaps, the burden and the privilege of our evolution.

Earth has been "designed" to exist for several billion more years, according to the normal life expectancy of the solar system. Whether that will happen, we don't know. But protecting our planet has an urgent moral and critical dimension to it and must become a global priority. Our theology, our biblical interpretation, and religious and ethical energy drawn from all spiritual perspectives should focus on the health and preservation of the planet.

We are emerging from the environmental Garden of Eden. We were relatively naïve in our understanding of the planet in previous centuries. We are awakening, and in the awakening we can see more clearly our needed role in the new Earth.

In summation, each generation chooses to engage challenging issues to the best of their collective ability. I addressed only three social problems, the issues of capital punishment, stem cell research and a growing perspective among many that Armageddon is near. These are just a few of the many social challenges we face, but they point out the necessity to embrace the best science we have along with the best faith. The faith that enabled these views comes from a belief that God encourages us to wrestle with social issues and gives us the freedom to do the best we can. Sometimes we will make mistakes but together faith and science can move us forward. The scalpels of science and faith-born ethics are needed to strengthen the dialogue that will enable us to make the best possible decisions.

A BRIDGE TO LIFE IN THE COSMOS

Be fruitful and multiply, abound on the earth and multiply in it.

—Genesis, 1:17

The more I examine the universe and the details of its architecture, the more evidence I find that the universe in some sense must have known we were coming. (133)

—Physicist Freeman Dyson

Genesis advises, "Be fruitful and multiply." (134) We have done that, some think, to almost the breaking point. The earth abounds with creatures. The human population could quite possibly double in the next fifty years. Barring catastrophic events—war, disease and natural calamity—we will, within a few centuries, discover that the planet Earth has grown too small for its occupants.

Before we reach that point, depleted resources will strain the global community. We will have to come up with many creative solutions. I can foresee a time when we mine the solar system for resources of every imaginable kind, a kind of space age gold rush. Within our solar system, the planets and their moons are a treasure trove of natural resources. Eventually, I can envision human colonies stretching across the Solar System and beyond, much as explorers in the 15th and 16th centuries crossed the seas and discovered new lands.

Both these potential challenges, the prospect of an overpopulated Earth and the need for more resources suggest to many that we should not abandon our interest in space or surrender our program entirely to other countries. We must remain a creative leader in this field, and should invite the corporate world into it so that as a global community we may find solutions to these challenges as they arise.

As we ponder these future realities, we are prompted to ask philosophical questions, such as: What is the meaning of our place in the cosmos? Is there life anywhere else? Is there life everywhere else? Are we connected to life everywhere? Are we destined to commune with the rest of the universe?

These somewhat distant concerns about the planet's future have taken some concrete forms in the past four or five decades. National pride has fueled much of our interest in space exploration. Our fledgling space program, which has since kindled similar programs in other countries, began being more competitive than cooperative. The climate of trust has improved with each decade since the 1970s. Ultimately, international cooperation is essential, enhancing the program with more minds working on the various challenges and less expense to individual countries. With the International Space Station, we see the fruits of this expansive philosophy. The scientific curiosity alive and well and the corporate faith in our shared future will bond us together, and a joint Earth-to-space adventure will not be denied. Space will become a significant Earth program. It may stumble and change directions and leadership, but eventually I believe it will be a lasting and important part of the human endeavor to reach beyond ourselves. From an American perspective, the adventure into space depends upon our corporate attitude toward space exploration and our will to persevere, since without the people's passion for exploration, governments will not be able to fund this expensive enterprise. Just as critically, the extent of international cooperation in space will be directly proportional to the degree of global civility and useful commerce we are able to experience with the world's nations. This "final frontier" will require the will of the people of planet Earth.

If ever there was an example of an enterprise that sits on the shoulders of the past giants of science and faith, it is launching the space program. Our debt to our forebears began many centuries ago—to at least the ancient Arabic students of the sky; and to Egyptian mathematicians such as Eratosthenes who, with amazing accuracy, measured the circumference of the earth in the second century B.C. (135) From giants of the Middle Ages such as Copernicus, Galileo, Brahe, Kepler, and Newton; to modern-day icons such as the father of rocketry, Wernher von Braun; to and deep space relativistic thinkers Einstein and Hubble, we humans have leapt light-years forward in our understanding of the universe and the forces that keep it together.

The space program is among the more glamorous fruits of their collective knowledge and work. It has spawned a tree of knowledge useful on terra firma. NASA claims more than fifteen hundred practical applications for human earthly use to which it has, wholly or in part, contributed. Under the Space Act of 1968, NASA has had a mandate to share publicly its discoveries including the following practical applications: GPS systems in our cars and boats, satellite weather maps, a portion of the technology in CAT scans and MRIs, heart defibrillator technology, remote robotic surgery, monitoring neutron activity in the brain, pay-at-the-pump satellite technology, cordless tools, airport baggage scanners, precision navigation, and freeze-dried food. (136)

Out of space exploration, we have formed new questions about the universe and our place in it. We have a little better idea of where we come from and some hints about our future. The program has rekindled an enthusiasm that with good reason suggests that we are not alone in the cosmos.

We have come a long way. It was not so long ago that all of this seemed like science fiction. I remember the sentiment that surrounded the acquisition of my first telescope in 1952. My father was worried that I was interested in so impractical a hobby. While he was pleased that I showed some interest in something academic, he held the prevailing view at the time that the stars were for poets, and the earth set our boundaries and limits. Many agreed with him and some still do.

A decade later, John F. Kennedy inspired and challenged America to leap into space, believing that America would be left behind if it did not pursue "the possible." On May 25, 1961, the president addressed Congress. He challenged his fellow Americans: it is "time for a great new American enterprise… time to take a clearly leading role in space achievement, which in many ways holds the key to our future on earth." The first specific challenge is the one that is most remembered.

> "First, I believe that this nation should commit itself to achieving the goal, before this decade is out, of landing a man on the moon and returning him safely to the earth. No single space project in this period will be more

> impressive to mankind, or more important for the long-range exploration of space; and none will be so difficult or expensive to accomplish." (137)

Pundits debated whether it was sensible or not. Scientists debated how to go about it, but the populace got excited and came aboard with the idea. The people's acceptance and excitement was critical, for they would, in the long run, be responsible for spending the money. As with most big projects, there was a political edge.

One of the large groups of people within the United States to be convinced of the worth of space exploration was "the people of faith." Millions of Americans had to assimilate the idea that money spent for a project outside Earth had substantial merit. And theologians, preachers and laypersons weighed in. If theology were to embrace the cosmos as well as the earth, people would be more likely to support programs that seek to explore other parts of God's universe. Very subtly we began to see ourselves as, perhaps, not the only viable planet in the universe, and began to dream that we might not have to be isolated from others forever. On the one hand, it was a more humble view, on the other, more extravagant.

Kennedy gave credence to the field of astronomy and the study of the universe. Quickly, other fields attached themselves to this adventure, and to complex space travel. A more aggressive study of the universe evolved. Science programs were energized.

Let us review the idea that life is abundant in the cosmos, beyond our Earth. Are the chances good or not?

Annie Dillard, the wonderful naturalist author, once wrote in her Pulitzer Prize winning book *Pilgrim at Tinker Creek* that the Creator, if nothing else, was prolific in creation. Many times more life forms have come and gone on this planet than exist today. Hundreds of thousands of species still exist. "After the one extravagant gesture of creation in the first place, the universe has continued to deal exclusively in extravagances, flinging intricacies and colossi down aeons of emptiness, heaping profusions on profligacies with ever-fresh vigor."(138) That makes it even more amazing when one considers the number of species that have crawled or walked or flown or swam on the earth and in the seas since the beginning.

Life existed, of course, for millions of years before human beings were much of a gleam in the Creator's eye. It is reasonable to assume that an intelligent Creator would celebrate and embrace the manifold creatures of the planet. Not all life is important only in relationship to our own. Given the plentiful and different forms of life upon this planet, it does not seem to be much of a reach to postulate that in the universe there is abundantly more.

Carl Sagan, in his acclaimed book *Cosmos*, once postulated that by the most conservative estimates, there were millions of extant civilizations in our galaxy alone, and tens of trillions of places in the universe where life might emerge. (139) Using common sense, it seems beyond reason to exist in a universe where there are billions of galaxies and only one of them supports life.

It is an appetite-whetting possibility that some forms of life outside the earth exist, even in our solar system, and there is evidence to support this idea. First, as has been mentioned, the simplest life forms existed here for a billion years before human beings arrived. So, life does not depend upon us. It also suggests that during that time we were not the be-all and end-all of creation. Second, for years it seemed that within our own solar system, only Earth sustained life. Yet we now know that many of our solar system's moons may have water and/or ice and some have hot inner cores. We also have some evidence to support the idea that some of the building blocks for life have come from meteorites, suggesting that life may be elsewhere within our solar system.

Moreover, Robert Ballard, whose submersible vessel first investigated the Titanic, also discovered forms of life deep in the ocean that do not owe their origin to life on the surface but rather to deep volcanic activity miles below the surface. Although all bodies in the solar system came from the sun, sunlight has no apparent role in their origin or in sustaining their life. This insight opens up even more possibilities for life on the planets than we had imagined only a few years ago.

I enthusiastically join with those who think that life is part of the fabric of creation in the universe. It leads me to inquire, how will people of faith adjust to this probable theme of extravagant existence throughout the universe? That is the subject of the next chapter.

AN EXTRAVAGANT CREATOR: A SPIRITUAL PERSPECTIVE

The heavens bear witness to your wonders, O Lord, and
to your faithfulness in the assembly of the holy ones;
For who in the skies can be compared to the Lord?

—Psalm 89:5-6

The whole history of science has been the gradual realization that events do not happen in an arbitrary manner, but that they reflect a certain underlying order, which may or may not be divinely inspired. (140)

—Stephen W. Hawking

Because it is probable that the universe is teeming with life, this chapter is about the need for a spiritual attitude adjustment. We may now regard the God in whom so many have believed through the centuries as an extravagant Creator. Expanding one's vision of the Cosmos expands one's vision of God and alters one's view of humanity. The God in whom Jews, Christians, Muslims and others have believed has a much vaster arena in which to work and perhaps much more to do than we ever imagined. God is the Lord, we believe, of countless worlds, not merely our own.

That is where the attitude adjustment must begin. It means that the universe is not principally about Earth's human beings or even all of Earth's creatures. In this light, how may people of faith regard themselves? If we are not the principal characters, then who are we? What is our purpose and our role?

In the New Testament, Jesus uses the term steward to describe good managers. In Matthew, a steward pays the laborers in a way that symbolizes heavenly justice. (141) In Luke, he tells of a steward that has to make up for bad management. (142) St. Paul, in his letter to Titus, says that bishops must be blameless stewards. (143) Steward is an honorable title that in Christian parlance describes our roles on Earth—we are to be stewards of the planet, stewards of our communities and of ourselves. We must manage ourselves well, and live by honorable ethics. This is in distinction from being owners, where we set the rules and others abide by our wishes. As stewards, we work with the understanding that we have a heavenly boss.

This philosophy is most evident in Jewish Temples and Christian Churches in regard to their mission goals. Stewardship campaigns rely on the belief that we are caretakers, not only of our own congregations, but also of the community and the world. From giving aid when widespread disasters occur to supplying food to local soup kitchens, religious institutions regularly respond to various needs. The ancient word tithe or tenth that dates back to the patriarch Abraham (144) is a giving model that is once again a commonplace standard today.

It makes a difference what we call ourselves, because our title reflects our purpose. Just as people of faith may identify themselves as stewards, scientists also have found an attractive appellation. They enjoy calling us star children, which arrests our attention. Star children is an apt name because all the molecules that make up our bodies were generated in the sun long before the earth existed. This name contains a truth that is important to digest. All of life comes from the stars.

People of faith believe that the spark that produced life from inorganic matter was the Creator's doing. They also believe that the stars themselves came into being at the Creator's behest.

Both the faith and scientific terms steward and star children suggest what our role should be in the future exploration of other worlds. As star children, we should be able to identify with other star children, even though they may look and act quiet differently from us. They, too, will be made from the stuff of stars albeit configured in different ways.

When we have developed the capacity and have the desire and will to explore other worlds, our role as stewards will be vital. Should there come a time when we communicate with other sentient beings, we cannot approach them with a conqueror's ideology and agenda. It must be as if it were a meeting of long lost cousins, both creatures of the stars and of the spirit. This way of treating other beings takes practice and is one more reason to begin practicing now.

It is quite likely that we will begin to use our solar system for its resources long before we meet and communicate with intelligent life elsewhere. Garnering such resources from the solar system is within our grasp in the next fifty years or so. While it may be that we find life in the solar system within the next century, it would be a startling surprise to find intelligent life. In all probability, our dream of communicating with intelligent beings will have to wait until human technology leaps forward greatly. But since the dream must precede the reality, I offer these thoughts on what we might find and how we should proceed from a faith-based perspective.

On Earth, there are many thousands of species, and they take remarkably different forms. Yet we recognize essential likenesses; eyes, ears, noses, hair, mouths, arms and legs, etc. Intelligence is recognizable as well. Elsewhere in the universe, other star children may have developed some of these "essentials." But, more assuredly we would recognize intelligence as in the manifestations of music, art, math, and language. We might even recognize more emotional states as anger, friendliness, caution, fear and affection. They may also have spiritual perspectives and may be attracted to faith in a Supreme Being or have another way of experiencing awe at their surroundings.

In Maria Doria Russell's fictional book, *The Sparrow*, she posits that the meeting between the species from another planet and we humans from Earth may not go swimmingly, and that much confusion will accompany our attempts at communication. (145) That may well be, but while her book raises prudent questions and suggests caution, her story is set only a few years in the future. Hopefully when the time really does arrive, we will be very well equipped, socially, scientifically and spiritually to make meaningful contact.

Here is the upside of what we may discover. There would be a worldwide sense of excitement in verifying that we are not alone in the universe. Imagine how people of earth would react. In every land, people would discuss its meaning and debate its implications. International forums would ponder how to continue the connections. Earth's people would have a common goal.

From a faith perspective, we would rejoice in the confirmed knowledge that the Creator is indeed splendid in His multiple creations. We would begin to look at ourselves less parochially. We would quickly acknowledge that any one of our philosophies or religions or outlooks on life is not the be all and end all of worldviews. We would try to find the similarities among us and affirm the values in our world that seem to hold true with our new friends from afar. We would begin to see our earthly community as one rather than so divided as we are now, and perhaps we would be more appreciative of the beauty we find here. We would appreciate more just how well suited we are to our own Earth and discover that, while other places hold intriguing possibilities for adventure and maybe even for resources and friendship, that as Dorothy in the Wizard of Oz finally discovered, there is no place like home. Billions of years of evolution here have given us, perhaps, "the best of all worlds."

For people of faith, connections to others outside our planet would encourage us to find value in the religious experiences of others, and we might enjoy sharing the saving graces of our own faiths along with the saving graces of their faiths. These are dreams of the distant future, but just as the dream must precede the reality, the vision and the hope must precede the glory of it all.

Now, as a believing Christian and an ecumenist who believes that God loves and saves through other faiths as well as my own, what does the prospect of life out there say about our belief in the Lord Incarnate? For me, God manifesting Himself on this planet does not preclude God manifesting Himself on some other orbs in the cosmos. In fact, I think it would be cruel and ungodly in character, not to do so. God is Spirit and can do as the Spirit wills. If God is Love, that Love knows no limits and certainly stretches across the cosmos. The earth itself has limits; it will not exist forever, but the Creator undoubtedly will. Therefore, it makes sound theological sense to believe that God's love is abundant here and also out there in the rest of the universe.

Therefore, we ought to rejoice in being part of such a grand design and, though our time is relatively short in cosmic terms, our spirit, as part of God's Spirit, is eternal. What a joy to be part of the eternal drama of life!

DARWIN AND JESUS, A FITTING MERGER

There's always a boundary underneath a bridge.
The internal bridges I build need the same transcendent quality
over the murky waters of my definitions. (146)

—Jacques Linder

It is incorrect to imagine that a particular species in time has unlimited evolutionary opportunities ahead of it...Homo sapiens was one of a range of possibilities in the evolution of the hominid group, not an inevitable product of that process. As long as we appreciate the true nature of our history, with its uncertainties, we will have a clearer sense of what it is to be a member of the species Homo sapiens. (147)

—Richard Leakey

It is a tale of two fish, one with legs and the other without. Which came first, Darwin's fish or the Christian fish? As with most ambiguities, the answer depends on how you look at it. As a symbol, the Christian fish has the edge. But, of course, the real squirmy fish predates all human life and consequently human symbols. The Christian fish has no legs. The Darwinian fish has legs. On bumper stickers and windshields across America, the two fish advertise the automobile owner's point of view—either God as author of creation, or evolution without the benefit of a deity. It is presumed the two fish represent opposite views that cannot be integrated. As you may imagine, I like cars that flaunt both fish.

In this final chapter, it is fitting to celebrate Darwin and Jesus together. This is not an insult to either figure; the two epical, historic figures are not in competition as so many seem to think. I imagine that Darwin and Jesus would not feel in competition with each other were they to coexist in the flesh at any period of history. However, we love competitions, and we create them even when no rivalry is obvious.

Whether the Christian fish symbol is on jewelry or stuck on cars, it sometimes displays strange Greek letters within the outline of the fish, and sometimes the outline alone tells the story. The legs protruding from Darwin's fish suggest that evolution trumps a biblical interpretation of Creation. Messages such as, "My God is not dead, sorry about yours," sometimes accompany the symbols to ratchet up the conflict. It appears that many people think that if one celebrates the Christian fish, one cannot celebrate Darwin's fish as well. In order to have a story to tell, we typically enjoy polarizing points of view rather than doing the work necessary to find their commonalities. It is easier to believe that Darwin and Jesus, science and faith, are mutually exclusive. Nothing could be further from the truth.

In the so-called competition between the two fish, the origin of the fish as a Christian symbol is often lost. Let us take a quick sojourn into its origins. The fish as a Christian symbol is derived from the early days of Christianity, before, in fact, the word "Christianity" existed. People who believed in the resurrection of Jesus called themselves followers of "The Way." (148) The Way was a path of living and believing that Jesus modeled. The early followers of Jesus were persuaded that through his resurrection, they, too, were alive eternally.

This fledgling group of believers was seen as a threat to the established religion of Rome. The Way was declared illegal because adherents did not bow before the emperor or worship Roman gods. These early believers were not allowed to congregate or to show any sign of their belief in a public setting. Instead, they were compelled to meet in secret. Those who chose to meet together anyway risked torture and imprisonment if they were discovered. Some paid the ultimate price.

The followers of The Way developed a secret code. It was a fish. "Fish" in Greek appears like this, IXTHUS. Each letter stands for part of the code. The code is an acrostic that stands for "Jesus Christ, Son of God Savior." Often the fish outline was drawn in the dirt outside a home where a simple worship service was about to take place. The service would have been a primitive Eucharist or an early Agape meal celebrating the fellowship of faith. The fish ultimately became a symbol of Christianity's dangerous beginnings. It called to mind the daring and courage it took to be a Christian in those early days. As in other movements, many people were impressed that these first followers believed so strongly and, as a result of their faith, The Way flourished.

The Darwinian fish is a symbol of one of the greatest scientific advances in human history. Darwin was as important as the earlier giants of the heavens, Copernicus and Galileo. These men convinced most of the world that the sun was the center of the solar system and that the planets, including the Earth, revolved around it. Darwin was an early champion of an Earth that had been here for millions of years, during which time the many species evolved. His view challenged the widely held belief that an omni-creative deity had made each of the species from scratch. The idea of a changing Earth and evolving life was counter to the basic world view that most educated people took for granted. To accept evolution would require a paradigm shift of nearly unprecedented proportions.

Charles Darwin began life in Shropshire, England, the son of a wealthy mother and a father not born to wealth but who, as a doctor, made an excellent living. In the seminal book *Evolution*, Carl Zimmer relates that young Charles spent his time in the country looking at bugs and beetles. He loved biology. He was not much interested in other academic pursuits. He was slated first to be a doctor, but hated medicine, and so his father thought the next best thing was for him to become a cleric. But at age twenty-one, Charles persuaded the twenty-seven-year-old skipper of HMS Beagle, Captain Robert FitzRoy, to let him be his scientific companion on his historic sail in 1831. FitzRoy's trip was to take him from Plymouth, England, down the east coast of South America, around Cape Horn and up the west coast of South America. The five-year mission included other tasks such as making detailed maps and testing clocks for the British navy.

Life for Charles Darwin changed dramatically following this epic trip that peaked in the leg of the journey that included the Galapagos Islands. His grand theory did not form in his mind all at once; the ideas that hatched in the Galapagos took years to coalesce and mature before any reputable biologist would pay attention to him. The journey on the Beagle with FitzRoy cemented in Charles's mind that he wanted to pursue biology. He also was now convinced that geology and biology should work hand-in-hand. After returning to England, he eschewed the life of a cleric that had been planned for him and began to work in earnest on developing his evolutionary ideas. Those ideas came to fruition in 1859 with the writing of his ground-breaking and ground-shaking opus, *The Origin of the Species*. (149)

The ground shook because his theory challenged the common understanding that the Deity had made the creatures on the earth just as we find them today—in a static, unchanging form. Evolutionary theory suggested, instead, that all creatures, including human beings, adapted to their environment or perished. This process had taken place over thousands if not millions of years, he suggested. Giraffes got longer necks; ostriches grew heavy, and though they kept their feathers, they lost their ability to fly; humans grew out of a family tree that included chimpanzees and, perhaps, gorillas and orangutans.

These notions were considered an outrageous insult to human beings and, if believed, presented a huge challenge to those who supported a literalist interpretation of biblical Creation. It may be difficult now to imagine just how angry people were, but when the metaphorical ground on which they stood shook seismically, passions were ignited. Darwin had reduced human beings to the level of an intelligent, swinging ape, or a dog or worse. Man was, after all, made in the image of God.

War ignited; it was a nineteenth century media frenzy. A withering attack on Darwin ensued. Many otherwise intelligent folks thought he was depraved. Cartoons portraying him in league with the devil decorated newspapers. Church sermons vilified him. Church of England Bishop Wilberforce blistered him at a scientific conference that Darwin, himself, declined to attend. The ground was shaking; if Darwin was right, how would life go on if people's core beliefs were shattered. (150)

It took some time for society to re-establish a sense of equilibrium. Theologians went back to their drawing boards; many expanded their theological and biblical vision; even biologists had to adjust to Darwin's theory of natural selection. The settling of the earth took a while, and in some places it is still settling.

As it turns out, Darwinian evolutionary theory fits rather nicely with a theology that includes biblical and spiritual evolution. The expanded view of the universe and good theology also fit comfortably together.

If we place the lives of Jesus and Darwin on the same examining table, we are likely to agree on at least one thing: the way of life that Jesus taught and the theory of evolution that Darwin offered were born and fleshed out under adversarial conditions. Each man pioneered a new vision and a new way of thinking. Both were singularly focused and disciplined, and exposing their thoughts to a wide audience required courage and sacrifice. Each advocate had to stand up against the authorities of his own day. Each paved the way for the future.

Jesus and Darwin aren't the only examples of dramatic pioneers in the laboratories of science or the field of faith. Other icons of science and faith echoed the ethics and courage of Jesus or the brilliance and determination of Darwin. I highlight Moses, the lawgiver, emancipator of the ancient Hebrews; Galileo, faithful while enduring the scorn and punishments of The Inquisition; Luther, "I am a prisoner in conscience to the Word of God. I cannot retract and I will not retract;" (151) Newton, a man of faith, consummate scientific genius known most for his physical laws of nature formulas; Schweitzer, combining theological brilliance with pastoral and medical compassion; Gandhi, who inspired and sacrificed himself for his people. Thus, one can imagine that, were these pioneers to have lived at the same time, they would likely have felt deep respect for each other.

Darwin's work has had a huge impact on religious thought since the publication of *The Origin of the Species*. Theologians scrambled to understand his theory and to posit the questions: Was there a way to integrate science and theology, perhaps with a healthy, productive tension? Would greater understanding between the two disciplines allow people to appreciate both the God of Creation and the theory of evolution? Initially, neither side thought there could be much common ground, as human beings are wont to do. Each side castigated the other. But time has a way of unlocking prejudices, easing the pain and enabling good science and good theology to prevail.

Slowly, some theologians began to see that Darwinian thought opened up theology to new riches. The ancient biblical myths still had their important stories to tell, so rich in purpose for human life, but no longer could they be understood as offering an exact chronology to existence or a blue print for how life unfolded on the planet. Considering that the writers of the first chapter of Genesis had no scientific basis for their early myths, they came remarkably close to depicting the evolutionary story that Darwin revealed through his scientific undertaking. In the biblical Creation story, after the earth was formed, God began creating life on it. Then came vegetation; on subsequent days, creatures were created to abound in the seas, and then creeping things and wild animals on land and finally human beings emerged. If the six days of Creation can be seen as a metaphor of time, the process, or unfolding of life on this planet, is recognizable even today. Hats off to the writers of the fifth-century B.C. story of creation for the excellent mixture of faith and science of the time!

It is unfortunate that Darwin did not overlap in history with Joseph Campbell, the renowned twentieth-century American mythologist. Campbell has helped millions understand how powerful and insightful myths can be and, as he believed, actually reveal more deeply eternal human truths than a simple fact-based story can do. His insights could have assuaged many theological fears. Author of the four volumes, *The Masks of God*, a great study of mythologies around the world, Campbell confirms that myths express "the unity of the race of man, not only in its biology but also in is spiritual history." (152) As Richard Leakey wrote in Origins Revisited, "A people's mythology is its means of coping with the urge to understand the world, for mythology is a body of explanation, an embodiment of the Truth." (153)

Darwin's ultimate gift to religion is that he forced Bible scholars to reexamine their worldview. Theologians searched for a more comprehensive way of expressing the Creation story. Many who had, perhaps, slipped too easily through difficulties of the six-day creation and the stories that followed, began to grapple with a new perception—the power of myth. During the following century, with the enthusiasm engendered by scholars using historical criticism, mentioned earlier, the riches of the Bible would be opened up anew for many readers. For those who realized that literalism was inadequate, the power of myth released a new appreciation of the ancient stories.

An important example of one of these that begged a more up-to-date interpretation was the story of Noah and the great flood. That there was a great flood, there is little doubt. We are now able to appreciate that the biblical rendition of the great flood was written about by earlier cultures as in the old Mesopotamian *Epic of Gilgamesh*, which predates Noah by at least five hundred years. (154) Robert Ballard, in his investigations of the Black Sea, suggests strongly an even earlier date for the flood—as long ago as 5600 B.C. (155) Darwin, perhaps unwittingly, was a catalyst for this paradigmatic shift in our thinking. Much of the world is still adjusting to this new evidence.

Darwin, in true evolutionary fashion, provided theologians who wanted to go beyond earlier understandings of Creation with much earthly data for investigating the biblical story and the creative process itself. Nineteenth century biblical scholars, especially German scholars, who demythologized much of Genesis through a process known as "form criticism," were the advance guard of this change in biblical perspective. Scholars such as Julius Wellhausen, Hermann Gunkel, Rudolf Bultmann and Martin Dibelius were giants in this field. (156) Darwin helped emancipate us from the rigidity that literalism fostered.

He gave us a new awareness of our biological origins and, more importantly, loosed a creative spirit among us. I rejoice in the gift, as do many students of the Bible and others from different religious persuasions. So, along with biologists and many scientists, we people of faith shout, "Viva Darwin!"

At the same time, the Bible, more than any other piece of literature, has formed my faith and enabled an ethical perspective for most of my day-to-day life. My faith and ethics have been fired in the biblical kiln. The issues of life and death, of love and hate, of pain and sorrow, of disillusion and reconciliation are grappled with in the poignant sagas that flow through the biblical pages. More than that, my reason for being, my sense of purpose, emerged and gained substance as a result of studying those rich testaments. And, it was the life of Jesus that impressed and captivated me the most.

Both Darwin and Jesus are heroes. One was a scientist who opened up a new way to understand the evolution of life on this planet; and the other, a simple teacher who opened up a new and courageous way to live life. Together, the two giants joining their hands across time can enable scientists and people of faith to rejoice at the top of the bridge overlooking the imaginary chasm.

Is there a meeting ground between science and faith, between Darwin and Jesus? My answer is an unreserved yes! Science and theology, curiosity and faith, have nothing to fear in engaging each other. Both disciplines are fueled by the aphorism, "The truth shall set you free." My hope is that science and faith may join hands and celebrate the different ways they have of making discoveries and human connections.

EPILOGUE

ON THE SEA OF GALILEE

In the first chapter, you will recall that, following their imaginary journey aboard the Beagle to the Galapagos Islands, Jesus and Charles Darwin planned a second engagement. The wrinkles in time once again made the meeting between the two men possible. This time in a milieu more familiar to Jesus, they met on the Sea of Galilee. For Jesus, the Sea of Galilee had been a refuge, a place to get away from the crowds, to meditate and to spend some time alone as well as provide an opportunity to commune with his chosen disciples. A carpenter by trade, he probably learned a good deal about fishing from them as well.

The "Sea" is really a freshwater lake that empties into the Jordan River. It flows some seventy miles to the Dead Sea, which is just fifteen miles to the east of Jerusalem. The large lake was, and still is, the lifeblood for many towns; fishing and allied industries provide work for much of the population. In the first century A.D., Jesus spent the majority of his ministry in those towns, including Capernaum. It was as close to an earthly home as Jesus had in his adult life. The disciple Peter's mother-in-law lived there, (157) and Jesus was an occasional houseguest. From there he preached to many local people and healed several individuals.

Capernaum, on the northwest shore, is in ruins today, but the Jesus of my fanciful tale, feeling a bit of nostalgia for the old town, wanted to begin the sailing adventure near his old haunts. So Charles Darwin and Jesus of Nazareth chose a clear afternoon with only a slight northeast wind to aid them and embarked near the shore of ancient Capernaum. They had engaged a simple, single-mast fishing boat that could be managed by two decent sailors. As they set sail they could see the low lying hills bathed in sunlight on the eastern shore not more than seven miles in the distance.

"It is good to be with you again," said Jesus.

"It certainly is," Charles responded. "I so enjoyed our conversation in the Galapagos and am eager to talk more about the concept of biological evolution and the evolution of the spirit. I am very curious about how humanity will resolve these ideas."

"My sentiments as well."

"One of the things I have been thinking about since last we met is what the consequences will be for mainstream theology if human beings are perceived to be only an offshoot of a lower primate species. If we share the same family tree as the apes, are we necessarily elevated above them? Are we intrinsically of more worth than they are? What does my theory of how the species evolve suggest about either the importance of human beings or, conversely, the importance of all other animal species?"

Charles paused a moment, but Jesus sensed that he had not quite finished his thought. Charles breathed in the fresh breezes off the Galilean lake and then resumed. "The Christian philosophy that I grew up with insists that man has a soul, and in this important respect is separated from the 'animals.' But what in the evolutionary path would have produced that distinction? How is it that we humans have a soul and animals do not? Have we just appropriated a soul for ourselves and denied it to those we call lower on the scale of life?" Charles paused again, but this time Jesus knew that the biologist would wait for a response.

"Yes indeed, those are important questions. In my day, we didn't address those questions so directly. As you remember we were just ending the practice of sacrificing animals on the altar. It is one of the reasons why later Christians sometimes referred to me as the lamb of God. But, you may remember reading in the Scriptures what I said back when I was preaching: 'Not even a sparrow falls without the Father's knowledge.'(158) We are all connected. We share a common spirit. But, because of our evolved mental abilities and spiritual attributes, particularly our consciousness of God, it means, of course, that we have additional responsibilities. This God-consciousness motivates us to take on the role of stewards of the planet as well as to take care of one another. We are our brothers' keepers. In time, humans will include the 'animal' kingdom in that concept."

"You speak about a common spirit for all creatures, but would you say that an animal has a soul?"

"Aha," Jesus smiled, and thought to himself for a moment, aware of the increasing number of churches that have a ceremony they call The Blessing of the Animals, "I strayed a bit from your question. I believe that God, the Father, cares about the animal kingdom in ways appropriate for them and for humans in ways appropriate to us. We all have God's Spirit within us. I speak of the spirit more than I do of the soul. I leave the distinction to the Greeks who were fond of the term *soul*. Let's talk about the spirit within us and among us. (159) Our consciousness makes us better equipped to develop that spirit. For example, compassion is not unknown in the animal kingdom, but it is a more highly evolved trait in humans. Compassion can blossom with sound ethical teaching. Compassion for all God's creatures is a potentially great gift from humans to the rest of creation."

"I hope so. I can relate to that point of view and it is not antithetical to what I have observed in the biological realm. On a biological level, I think we have a shared ancestry with animals and, while the fittest survive, the interconnections of the different species are observable. What we call compassion can easily be observed in the animal kingdom and occasionally bridges the different species. For example, at a wildlife watering hole we have observed that different species that are potential prey often warn one another when danger is looming. Occasionally a huge hippo, which is prey for no one, will charge a crocodile that has grabbed and dragged a delicate impala into the water and then, remarkably, attempt to assist the wounded impala."

"Many of the higher creatures in the animal kingdom have the capacity for tenderness and compassion. What separates humans from the rest of the animal kingdom is our capacity to be conscious of a divine presence. Because of it, we can engage God in ways that the others cannot. When we evolved to a point that we could reach for God, for the Divine in our midst, our spirit grew vastly."

"Are you suggesting that the spirit or soul within us can grow or evolve?" Charles asked.

"Yes, when we nurture the spirit with acts of love and care for others, we get better at it. It is not so different from exercising our muscles. Spiritual muscles need to work in order to develop. (160) I would also point out that while the planet's health was not so much an issue in my day in Israel, increasingly through the powers science has given to modern humans, we can injure and even desecrate the planet."

Charles rejoined, "One important possibility for humans is that in the future we will have the capacity to influence evolution rapidly in significant ways. Animals can only do that over a long period of time. How we will manage those changes responsibly and with wisdom will be one of the biggest questions of future ages. The altering of our humanity through science will have enormous ethical implications. I would add that animals also strive to survive, and this also has an effect on evolution in the long run. It is not the environment alone that produces mutations or changes. The animals engage with their environment and without their willingness to struggle, there would be little adaptation."

Both men were quiet for a minute or two. Then Jesus said, "You know Charles, one could easily make the case, from what you are saying, that humans owe nearly everything to the struggle that animals have gone through in the past; for without their struggles and adaptations, we wouldn't be human. It seems so simple, but we rarely look at animals and say thank you."

As their small sailing vessel drifted out into the middle of the lake, Jesus and Charles fell silent briefly and meditated on their conversation. The breezes blew and they looked out at the vistas. From where they were now they had a panoramic view of the northern and eastern shores of the Galilean lake; the view from this vantage point was very similar to the view Jesus enjoyed with his disciples years before. After a while, with a big sigh, Jesus came back to the present.

"Now in the twenty-first century, humanity teeters on a dangerous precipice. It is ironic that we meet here on this sea that was home to me during my earthly existence and now the surrounding area is a seething cauldron with global implications. It is particularly sad since religion is at the heart of the difficulties especially between Jews and Muslims but also affecting Christians. We should remember that those who live by the sword will die by the sword. Only the ancient admonition to love one's neighbor as oneself can bring peace to this area.

"It seems that human evolution has brought us to a critical point. The instinct to survive is producing new challenges. For much of human history, it has not seemed to be apparent. For thousands of years we relied on our survival instinct to save us. This is now clashing with our capacity, for example, to make weapons that can utterly destroy us. Which one of these strong impulses will ultimately become dominant and survive?"

"Yes, we do seem to be on a collision course, and there is no predetermined ending. But, Charles, one amazing thing about

God is the freedom He has given us to love one another or hate one another, to live in harmony or to destroy ourselves. We can call on Him for help and guidance, but He won't force the best answers upon us. He has that much respect for us. Humans have, at times, gone in some terrible directions and He hasn't stopped us. It is only when our consciousness of God's aspirations for us inspires us to goodness that we escape from our darker selves."

"A lot depends on us," Charles responded. "In your opinion, would God let the world be destroyed if we humans do not manage well our calling to be good stewards of Earth?"

"A big question. As you know, life on the planet has been nearly annihilated in the distant past, so we know another calamity is possible. The planet did survive, however, and rebuild itself, several times. All of life is difficult to extinguish. The earth regenerates and life keeps going. Humans are persistent. I would not doubt that we will find a way to survive and to thrive, but we must work at it, especially in the complex world in which we live. And Charles, above all, hold on to the belief that the Spirit of God cannot be vanquished. With prayerful guidance, together we can create the future."

"Jesus, when you walked and taught around Galilee, you emphasized what you called the 'Kingdom of God.' It seemed to be a kingdom where God is particularly present when we love one another or do justice for another. If this is so, do you think that this kingdom is part of our evolutionary heritage?"

"Well, yes because humans have the capacity to commune with God and act with compassion and do justice. Inviting people to walk in the 'kingdom of God' was at the heart of my message. As you have stated well, the kingdom of God is not a place but an action. It is being in the heart of where God is. God's invitation was and is to begin your walk now. It begins on earth and fills eternity as well. When I said in your recorded book, Luke's Gospel, that the Kingdom of God had come near you and again when I said the Kingdom of God was among you I desired that people know that the eternal kingdom begins now. God is present now as well as beyond the heavens and after we die. Secular kingdoms of the world rise and fall, but the Spirit always reintroduces God's justice and mercy and love in its many forms. The kingdom of which I so often spoke may fade at times when people revert to excessive self-interest or greed, but it always re-emerges when people love unselfishly and work for the betterment of others."

"That tendency seems to ebb and flow in society as if we are not quite sure in which direction we want to go. I suppose that is where faith comes in. We have to believe that we are getting better in order to get better."

Jesus said, "That is what faith is about. With it, the impossible becomes possible. The Kingdom of God is well established within the lives of many people, and because of it, justice and mercy will always find a way to engage and defeat evil in the world. In that sense, it is the pinnacle of evolutionary advance. As long as we enlist the Father's help and, in a sense, partner with Him, there is no force that can ultimately defeat the human spirit. You might say God's spirit gives humanity the edge to survive.

"Amen to that."

Jesus sighed, laid back and let the beauty of the sea enfold him; from out in the middle of the Sea of Galilee, he remembered the times so long ago when he and his closest friends used to fish and talk and ponder the future. He was so glad to have added Darwin to their company. As he looked skyward, he thought how divine it would be to take other trips in others venues with some of the profound thinkers of history such as Socrates, Leonardo and Lenin; perhaps, iconic religious figures like Buddha, Moses and Mohammad. Jesus was drifting into sleep but his last thoughts were of traveling with some people of profound courage. He thought of Rosa Parks and Nelson Mandela, and he wondered about you.

Charles awakened him briefly to help hoist the sail. When they were sailing peacefully toward the port, Jesus said, "Why don't we just relax now, enjoy the remainder of the day, and let the breezes of Galilee flow into our spirits?"

Charles rejoined, "Galapagos or Galilee, the wind eases the spirit into wonderful thoughts."

ENDNOTES

Ch. 1 Giants on the Beagle

1. Matthew 9:29.

2. Letter from Charles Darwin to Joseph Hooker, 1884, quoted by Carl Zimmer, *Evolution*, New York, HarperCollins Publishers, p. 37.

3. Carl Zimmer, *Evolution*, pp.16 -18

4. Matthew 7:12.

Ch. 2 Coming of Age in Science and Faith

5. Carl Sagan, *Cosmos*, New York, Ballantine Books 1980, p. xix.

6. Charles Foster, *Story of the Bible*, Philadelphia, Charles Foster Publishing, 1911.

7. Deuteronomy 6:4–5, Mark 12:29–30.

8. The Rt. Rev. A. Ervine Swift, Episcopal Bishop of Puerto Rico, 1951–1965, letters written between 1951–2001.

Ch. 3 In Science, Seeing Is Believing. In Faith, Believing Is Seeing.

9. Teilhard de Chardin, *Le Paradoxe Tranformiste*, 1925.

10. *Associated Press*, Friday September 4, 2009, Melanie Oudin, "Believe."

11. Mark 9:24.

12. Tom Jones, *Sky Walking*, First Smithonian Books, inprint of Harper Co,llins Publishers, 2007 p.125

Ch. 4 A Breath of Fresh Air Blows Through the Bible

13. Vatican librarian, 17th century, quoted in *Galileo's Daughter*, p. 65.

14. Richard Leakey, *Origins Reconsidered*, New York, Doubleday, 1992, pp. 8 and 78.

15. Ibid. p. 29.

16. *The Oxford Dictionary of the Christian Church*, third edition, editor, Dr. E. A. Livingstone, 2006, James Ussher, 1581–1656, Archbishop from 1625, p. 610.

17. *The Random House Encyclopedia, New Revised Edition*, New York, James Mitchell, editor-in-chief, 1983, p. 48.

18. http://www.wisegeek.com/what-is-creation-theory.htm May 24, 2010; young earth creationism asserts that only 10,000 years have passed since God created the earth. The Omphalos hypothesis theorizes that God created Earth

in its mature form. These theorists state that growth rings and other presumed evidence of the earth's age were placed here by God at the time of creation.

Ch. 5 In Celebration of Biblical Evolution

19. Teilhard de Chardin, Esquisse d'un univers personnel, quoted by Christopher Mooney in *Teilhard de Chardin and the Mystery of Christ*, New York, Harper and Row 1966 p. 42.

20. Genesis 1:1.

21. *The New Interpreters Bible*, Abingdon Press, 1994 ,Volume One, Commentary on Genesis, p. 340.

22. Genesis 1:2–3.

23. Genesis 1:20.

24. Genesis 1:1–31.

25. Matthew Fox, *Original Blessing*, Santa Fe, New Mexico, Bear and Company, fourth printing, 1984, p. 26.

26. Exodus 20:8.

27. http://www.bukisa.com/articles/94583_15_significant-law-codes-that-help-shapedhistory May 24, 2010.

28. Deuteronomy 6:5.

29. Leviticus 19:18.

30. Mark 12:31.

31. Leviticus 24:19.

32. Matthew 5:38.

33. Psalm 59:11.

34. Matthew 5:43.

35. Isaiah 9:6.

36. Isaiah 53:2.

Ch. 6 The Faith of Some Giants of Science

37. Ecclesiasticus (Sirach) 44:1.

38. Albert Einstein, February 13, 2010, http://en.wikipedia.org/wiki/Albert_Einstein

39. Thomas Arny, *Explorations*, New York, McGraw-Hill, third edition, 2004, pp. 46–47.

40. Britannica Concise Encyclopedia

41. www.answers.com/topic/sir-isaac-newton May 19, 2010

42. Beatrice Gormley, *Maria Mitchell: The Soul of an Astronomer*, Eerdmans Publishing, MI, 1995, p. 6.

43. Gormley, op. cit., p. ix.

44. Gormley, op. cit., p. 96.

45. Gormley, op. cit., p. x.

46. Rabbi Herbert S. Goldstein's interview with Albert Einstein, 1929, http://en.wikipedia.org/wiki/Albert_Einstein February 13, 2010.

47. Ibid. February 13, 2010.

48. William Paley, *Natural Theology, Evidences of the Existence and Attributes of the Deity collected from the Appearances of Nature*, 1802.

Ch. 7 In Celebration of the Faith of Science

49. Teilhard de Chardin, *The Phenomenon of Man*, New York, Harper & Brothers Publishers, 1959, p. 223.

50. Psalm 91:1. 19:1

51. Sagan, op. cit., p. 3.

52. Eric John, *The Popes, A Concise Biographical History*, New York, Hawthorn Books, Inc. 1964, p.378.

53. Theodor H. Gaster, *The Dead Sea Scriptures*, New York, Doubleday Anchor Books, 1956, p. 33.

54. http://www.intentblog.com/archives/2007/02hqr_albert_eins.html February 13, 2010.

55. http://www.newworldencyclopedia.org/entry/Marie_Curie May 25, 2010.

56. http://www.raimon-panikkar.org/english/laudatio.html May 25, 2010.

Ch. 8 The Bridge of Light Reveals the God of Wonder

57. http://www.intentblog.com/archives/2007/02/hqr_albert_eins.html February 13, 2010. "Science: Philosophy and Religion: A Symposium," 1941.

58. Genesis 1:3–4.

59. Arny, op. cit., p. 96-97.

60. Stephen W. Hawking, *A Brief History of Time*, Toronto, New York, London, Sydney, Auckland, Bantam Books, 1988, p. 115.

61. John 1:3.

62. *The Interpreters Dictionary of the Bible*, Abingdon Press, Nashville, 1962, p. 130.

Ch. 9 The Spirit of God

63. Paul Tillich, *The Systematic Theology of Paul Tillich, A Review and Analysis*, Alexander J. McKelway, Richmond, Virginia, John Knox Press, p. 126.

64. John 4:24.

65. Tillich, op. cit., p. 114.

66. Matthew 6:9.

67. I Kings 19:12 (KJV)

68. Kahlil Gibran, *The Prophet*, New York, Alfred A. Knopf, publisher, 1961, p. 60.

69. E. L. Allen, *Kierkegaard: His Life and Thought*, New York, Harper & Brothers, Publishers, 1935, p. 198.

Ch. 10 An Integration of Science and Faith

70. Carl Sagan, *Cosmos*, New York, Ballantine Books, 1980, p. 275.

71. Ibid. p. 161.

72. Seth Shostak, *Boldly Going Nowhere*, The New York Times, OP-ED, April 14, 2009.

73. Thomas T. Arny, *Explorations*, New York, McGraw-Hill, 2004, p. 468. The orbital speed of the sun in the galaxy is 220 kilometers per second or 130 miles per second or 468,000 miles per hour.

74. Sagan, op. cit., p. 2. The speed of light: usually rendered at 186,000 miles per second or six trillion miles in one year, a light year. If reduced to one hour, light would travel about 669 million miles per hour.

Ch. 11 Earth's Place in the Great Scheme of Things

75. George Smoot and Keay Davidson, *Wrinkles in Time*, New York, William Morrow and Company, Inc., 1993, p. 296.

76. Arny, op. cit., p. 444.

77. Ibid. p. 331.

78. Wikipedia, http://wikipedia.org/wiki/Goldilocks_planet May 20, 2010. "A Goldilocks planet is a planet that falls within a star's habitable zone."

79. David H. Levy, *Skywatching*, Time Life Books, editor John O'Byrne, 1994, p. 263.

80. Arny, op. cit., p. 346.

81. Arny, op. cit., p. 311–312.

Ch. 12 Miracles of Faith and Science

82. Paraphrase of Psalm 19:16, by Joseph P. Addison, 1712, taken from the *Episcopal Hymnal*, Church Publishing, New York, 1982, p. 490.

83. Father Brown Omnibus, from *The Quotable Chesterton*, San Francisco, Ignatius Press, editors, Marlin, Rabatin & Swan, 1986 p. 218.

Ch. 13 Miracles Close to Home

84. Father Brown Omnibus, loc. cit., p. 218.

85. Barbie Riegel, email letter, September 28, 1999..

86. Luke 8:48.

Ch. 14 Traces of Arrogance in Science and Faith

87. Teilhard de Chardin, quoted in Christopher Mooney's *Tielhard de Chardin and the Mystery of Christ*, New York, Harper & Row Publishers, third edition, 1966, p. 60.

88. Arny, op. cit., p. 332–333.

89. Genesis 32:22–28.

90. John 8:7.

91. George Arthur Buttrick, Dictionary Editor, *The Interpreter's Dictionary of the Bible*, Abingdon Press, New York, 1962, entitled, Hammurabi, p. 517.

92. Lex talionis, known in Roman jurisprudence, *The New Interpreters Bible*, Nashville, Abingdon Press, 1994, p.1165.

93. Leviticus 24:19.

94. Proverbs 25:21.

95. Romans 12:20.

96. Matthew 5:39 and 5:44.

97. Acts 20:35.

98. Amos 5:24.

Ch. 15. A Bridge over the Troubled Waters of Pain and Suffering

99. http://www.qotd.org/search.html?aid=2855 May 27, 2010

100. http://www.coolquotes.com/categories/aristotle.html May 26, 2010.

101. Rabbi Harold Kushner, *When Bad Things Happen to Good People*, New York, Anchor Books, Random House, 1981, pp. 35, 90.

102. Mary Baker Eddy, *Science and Health, With Key to the Scriptures*, published by The Trustees under the Will of Mary Baker G. Eddy, Boston, U.S.A. 1875, p. 228–231.

103. Jolande Jacobi, *The Psychology of C.G. Jung*, New Haven and London, Yale University Press, sixth edition, 1962, p. 124.

104. Matthew 5:3, 11.

105. Doug Single, *Skill, Courage, Sharing, Joy—The Stories of Special Olympics*, Special Olympics International, 1992, p. 5.

106. Ibid., p. 5

107. Pastor Marcia King, St. George's Episcopal Church, Nashville, Tennessee, bulletin, May 11, 2010.

108. Courage to Change, *One Day at a Time in Al-Anon II*, Al-Anon Family Group Headquarters, Inc., New York, p. 367.

Ch. 16 A Bridge for Medicine and Faith

109. Albert Schweitzer, *Out of My Life and Thought*, an autobiography, first published by Henry Holt, New York, 1933; thirteenth edition, The New American Library, 1964, p. 77.

110. Matthew 9:22.

111. Luke 8:48.

112. The Second Book of Kings 5:1-14.

113. John 5:1–9.

114. *The Book of Common Prayer, Ministration to the Sick,* New York, Church Publishing Inc., 1979, edition, 1986, pp. 453–457.

115. BCP, op. cit., 456.

116. The Shrine of the Master, Sarasota, Florida.

117. Schweitzer, op. cit., p. 69.

118. Schweitzer, op. cit., p. 77.

Ch. 17 Bridges for Life's Issues

119. Jacques Linder, Sarasota, Florida, artist/poet, quote attached to a photograph entitled, Building Bridges, Siesta Key, Florida, Siesta Key, Florida 1996

120. Albert Einstein, Quotes.net.stands4LLC, http://www.quotes.net/quote/9410 May 28, 2010.

121. Genesis 9:5, Exodus 21:23, Deuteronomy 19:21.

122. Leviticus 24:17.

123. Deuteronomy 32:35.

124. Romans 12:19 and 12:21.

125. Exodus 20:13.

126. "…is paved with good intentions." Statement attributed to Saint Bernard of Clairvaux, 12th century A.D.

127. Jodi Picoult, *My Sister's Keeper*, New York, Washington Square Press, 2004, pp.350-373.

128. *The Interpreters Dictionary of the Bible*, op. cit., pp. 226-227.

129. Mark 13:32.

130. Jane Fletcher Geniesse, *American Priestess*, New York, Doubleday, 2008, pp. 3, 37ff.

131. Geniesse, op. cit., p. 310.

132. Carl Zimmer, *Evolution*, op. cit., pp. 161–163.

Ch. 18 A Bridge to Life in the Cosmos

133. Quoted from *The Anthropic Cosmological Principle* (1986) by John D. Barrow and Frank J. Tipler, p. 318 http://en.wikiquote.org/wikiFreemanDyson May 29, 2010.

134. Genesis 1:28.

135. Arny, op. cit., pp. 40-41.

136. http://spaceexploration.suite101.com/article.cfm/practical_applications_of_space_technology

137. JFK's address before Congress, May 25, 1961.

138. Annie Dillard, *Pilgrim at Tinker Creek*, New York, Harper Perennial Publishers, 1974. Modern Classics Edition, p. 11.

139. Sagan, op. cit., pp. 247–252.

Ch. 19 An Extravagant Creator—A Spiritual Perspective

140. Hawking, op. cit., p. 122.

141. Matthew 20:8.

142. Luke 16:1–8.

143. The Letter of Paul to Titus 1:7.

144. Genesis 14:20.

145. Maria Doria Russell, *The Sparrow*, New York, Villard, 1998.

Ch. 20 Darwin and Jesus, A Fitting Merger

146. Jacques Linder, Sarasota, Florida, artist and poet. 2010.

147. Owen Chadwick, *The Reformation*, Great Britain, Penguin Books, reprinted in 1990, p 56.

148. Richard Leakey, *Origins Reconsidered*, New York, Doubleday, 1992, pp. 347 – 348.

149. Professor Clarence Tucker Craig, Oberlin, *The Beginnings of Christianity*, New York, Abingdon Press, 1943, p.152.

150. Zimmer, op. cit., pp. 7-8.

151. Ibid. pp. 50–52.

152. Joseph Campbell, quoted in *Leakey's Origins Reconsidered*, p.306.

153. Leakey, op. cit., p.306.

154. Ian Wilson, *Before the Flood*, St. Martin's Griffin, New York, 2001, pp.19–23.

155. Robert Ballard, *Oceanographer and Explorer, the Titanic*, quoted in Ian Wilson's Before the Flood, pp. 59–65.

156. Rudolf Bultmann and Karl Kundsin, *Form Criticism*, New York, First Harper Torchbook edition published 1962, pp. viii

Epilogue

157. Mark 1:21-30.

158. Matthew 10:29

159. Luke 17:21

160. Ephesians 6:10-17 and I Corinthians 3:1-9

BIBLIOGRAPHY

Allen, E. L. *Kierkegaard and His Life and Thought*. New York and London: Harper and Brothers, 1935.

Arny, Thomas T. *Exploration*. New York: McGraw-Hill, 2004.

Bultmann, Rudolf and Kundsin, Karl. *Form Criticism*. New York: Harper Torchbooks, first edition, 1962; original 1934 by Willet, Clark & Company.

Chadwick, Owen. *The Reformation*. Great Britain: Penguin Books, reprinted 1990

Craig, Clarence Tucker. *The Beginning of Christianity*. New York and Nashville: Abingdon Press, 1943.

Dillard, Annie. *Pilgrim at Tinker Creek*. New York: Harper Perennial Publishers, 1974. Classics Edition, 2007.

Eddy, Mary Baker. *Science and Health, With Key to the Scriptures*. Boston: Published by the Trustees under the Will of Mary Baker G. Eddy, 1875.

Foster, Charles. *Story of The Bible*. Philadelphia: Charles Foster Publishing, 1911.

Fox, Matthew(.) *Original Blessing*. Santa Fe, New Mexico: Bear & Company, 1983.

Gaster, Theodor H. *The Dead Sea Scriptures*. Garden City, New York: Doubleday, 1956.

Geniesse, Jane Fletcher. *American Priestess*. New York, London, Toronto, Sydney, Auckland: Doubleday, 2008.

Gibran, Kahlil. *The Prophet*. New York: Alfred A. Knopf, 1961; first published 1923.

Gormley, Beatrice. *Maria Mitchell—The Soul of an Astronomer*. Grand Rapids, Michigan and Cambridge, U.K.: Eerdmans Books for Young Readers, 2004.

Hawking, Stephen. *A Brief History of Time*. Toronto, New York, London, Sydney, Auckland: Bantam Books, 1988.

Jacobi, Jolande. *The Psychology of C. G. Jung*. New Haven and London: Yale University Press, sixth edition, 1962; first published in England, 1942.

Jones, Tom. *Sky Walking, An Astronaut's Memoir*. Smithonian Books paperback ed. 2007

Kushner, Harold. *When Bad Things Happen to Good People*. New York: Anchor Books, a division of Random House, 1981.

Leakey, Richard. *Origins Reconsidered*. New York: Doubleday, 1992.

Levy, David. *Skywatching*. San Francisco and Sydney: Time-Life Books, 1994.

McKelway, Alexander J. *The Systematic Theology of Paul Tillich*. Richmond, Virginia: John Knox Press, 1964.

Mooney, Christopher. *Teilhard de Chardin and the Mystery of Christ*. New York: Harper and Row, 1964.

Paley, William. *Natural Theology, Evidences of the Existence and Attributes of the Deity collected from the Appearance of Nature*. London: 1802.

Picoult, Jodi. *My Sister's Keeper*. New York: Washington Square Press, 2004

Sagan, Carl. *Cosmos*. New York: Ballantine Books, 1980.

Schweitzer, Albert. *Out Of My Life And Thought*. New York: Henry Holt and Company, Inc., Published as a Mentor Book, 1953.

Single, Doug. *Skill, Courge, Sharing, Joy—The Stories of Special Olympics*. Special Olympics International, 1992

Smoot, George. *Wrinkles in Time*. New York: William Morrow and Company, Inc., 1993.

Sobel, Dava. *Galileo's Daughter*. New York: Penguin Books, 2000.

Teilhard de Chardin, Pierre. *The Phenomenon of Man*. New York: Harper and Brothers Publishers, English translation, by Julian Huxley, 1959.

Wilson, Ian. *Before the Flood*. New York: St. Martin's Griffin, 2001.

Zimmer, Carl. *Evolution*. New York: HarperCollins Publishers Inc., 2001.

Reference Books

The Concise Oxford Dictionary of the Christian Church, editor Dr. E. A. Livingstone, Oxford: Oxford University Press, 2006; first published 1977.

Courage to Change, One Day at a Time in Al-Anon II. New York: Al-Anon Family Group Headquarters Inc., 1992.

The Hymnal, 1982, New York: Episcopal Church Publishing Incorporated, 1982.

The HarperCollins Study Bible, (NRSV). editor, Wayne A. Meeks. New York: HarperCollins. 1993.

The New Interpreters Bible. New York and Nashville: editors, George A. Buttrick, Walter R. Bowie, Paul Scherer, John Knox, Samuel Terrien, Nolan B. Harmon, Abingdon Press, 1953.

The Interpreter's Dictionary of the Bible. Nashville: Abingdon Press, Dictionary editor, George Arthur Buttrick, 1962, 23rd printing, 2000.

The Quotable Chesterton. San Francisco: Ignatius Press, editors, George J. Marlin, Richard P. Rabatin, John L. Swan, 1986.

The Popes, A Concise Biographical History. New York: Hawthorn Books, Inc., editor, Eric John, 1964.

The Random House Encyclopedia. New York: Random House, editor, James Mitchell, 1983.

Newspapers

The Herald Tribune, Sarasota, C Section, The Associated Press, September 4, 2009.

The New York Times, *Science Times*, "Darwin," February 10, 2009.

The New York Times, *Science Times*, "In a Lonely Cosmos, a Hunt for Worlds Like Ours," March 3, 2009.

The New York Times, OP-ed page, "Boldly Going Nowhere," April 14, 2009.

Periodicals

Katherine Bourzac. *Technology Review*, MIT News, What We Don't Know in Physics, January/February 2009.

Letters

Letters from The Rt. Rev. Bishop A. Ervine Swift, (from 1956–2000) Bishop of Puerto Rico, 1951–1964.

Letter from Barbie Riegel, email, The Bermuda Dolphins, September 28, 1999.

Internet

Aristotle: http://www.coolquotes.com/categories/aristotle.html May 27, 2010.

Britannica Concise Encyclopedia, www.answers.com/topic/sir-isaac-newton May 19, 2010.

Albert Einstein, Intentblog: http://www.intentblog.com/archives/2007/02hqr_albert_cins.html May 25, 2010.

Einstein:

http://www.quotes.net/quote/9410 May 28, 2010.

Code of Hammurabi: http://www.bukisa.com/articles/94583_15_significant-law-codes-that-help-shapedhistory May 24, 2010.

Helen Keller: http://www.qotd.org/search.html?aid=2855 May 27, 2010.

John F. Kennedy, address to Congress, May 25, 1961. http://jfk/urgent+national+needs+page+4.htm May 28, 2010.

New World Encyclopedia: http://www.newworldencyclopedia.org/entry/Marie_Curie

May 25, 2010.

Practical space technologies: http://spaceexploration.suite101.com/article.cfm/practical_applications_of_space_technology May 25, 2010.

Raimon Pannikkar: http://www.raimon-panikkar.org/english/laudatio.html May 25, 2010.

Wikipedia: http://en.wikipedia.org/wiki/Albert_Einstein February 13, 2010.

http://wikipedia.org/wiki/Goldilocks_planet May 20, 2010.

http://wikiquote.org/wiki/FreemanDyson May 28, 2010.

Wisegeek: http://www.wisegeek.com/what-is-creation-theory.htm May 24, 2010.